SpringerBriefs in Latin American Studies

SpringerBriefs in Latin American Studies promotes quality scientific research focusing on Latin American countries. The series accepts disciplinary and interdisciplinary titles related to geographical, environmental, cultural, economic, political and urban research dedicated to Latin America. The Series will publish compact volumes (50 to 125 pages) refereed by a region or country expert specialized in Latin American studies. We offer fast publication time of 8 to 12 weeks after acceptance. The series aims to raise the profile of Latin American studies, showcasing important works developed focusing on the region. It is aimed at researchers, students, and everyone interested in Latin American topics.

More information about this series at http://www.springer.com/series/14332

Adam S. Dohrenwend

Green Gold

Contested Meanings and Socio-Environmental Change in Argentine Yerba Mate Cultivation

Adam S. Dohrenwend
Department of Geography and Anthropology
Louisiana State University
Baton Rouge, LA, USA

ISSN 2366-763X ISSN 2366-7648 (electronic)
SpringerBriefs in Latin American Studies
ISBN 978-3-030-82010-7 ISBN 978-3-030-82011-4 (eBook)
https://doi.org/10.1007/978-3-030-82011-4

This Springer imprint is published by the registered company Springer Nature Switzerland AG
The registered company address is: Gewerbestrasse 11, 6330 Cham, Switzerland

Preface

Today's globalized economy separates consumers from the consequences of their actions. Evaluating the impact of consumerism from a political-ecological lens, informed by the work of geographers Robert Sack and Ian Cook, can help guide an analysis that geographically reconstructs supply chains and reveals the realities of consumption. This text applies this approach to study the externalization of cost under capitalism in the production of Argentine yerba mate, an infusion with stimulant properties long-used by Indigenous peoples. The use of yerba mate has become a cornerstone of Argentine society and identity, and yerba mate processors are working to expand exports globally. In Argentina's Misiones province, the heart of yerba mate production, the true costs of production are borne by the children, the impoverished laborers, and the environment of Argentina's Atlantic Rainforest. This compact volume examines these consequences of modernity, along with the efforts of an NGO to remedy them.

Baton Rouge, USA — Adam S. Dohrenwend

Preface

Contents

Chapter 1
Introduction

Abstract This chapter introduces the text's subject matter, yerba mate, as well as its theoretical and methodological approaches. In today's globalized economy, the consequences of consumption are often unknown to consumers—in part, obscured by long distances between supply chain nodes and end product consumers. By utilizing a political-ecological lens informed by the work of geographers Robert Sack and Ian Cook, as shown in this chapter and the ones that follow, researchers can better reconstruct complex global supply chains and make for more conscious consumers.

Keywords Political ecology · Robert Sack · Yerba mate · Globalization

Yerba mate, a cornerstone of Argentine culture for centuries, has been thrust on the global commodification stage in recent years. As the vast majority of Argentines already consume the caffeinated infusion, the industry seeks a new global market through which it can increase profits. In order to accomplish this, a context has been constructed and harnessed to define both the market and the product. This context seemingly defines yerba mate by its benefits to the consumer in mind, body, and spirit; however, as is often the case, this narrative does not completely match reality. Today's global economic landscape, riddled with complex production chains, obscures social and environmental realities driven by consumption and makes the assessment of marketed contexts much more challenging. This is not unique to yerba mate. At a national supermarket chain, try asking the butcher where exactly the ground turkey or pork chops come from. Where do the shrimp come from? What about the corn that makes up the bulk of the cereal aisle? They may not even know the country of origin, let alone the local contexts of productions.

As consumers become more disconnected from the increasingly complex networks of hyper-specialized places of production that fulfill their desires of consumption, understanding the consequences of one's consumption can become seemingly impossible. When one consumes in today's economy, they enter a vast network of relationships, both with the people and the environments of countless places. Furthermore, to achieve low consumer prices at the end of a production chain, these often-obscured relationships can become increasingly problematic. This represents the externalization of cost onto producers, their neighbors, and the environment.

A. S. Dohrenwend, *Green Gold*, SpringerBriefs in Latin American Studies,
https://doi.org/10.1007/978-3-030-82011-4_1

These processes are not unique to yerba mate, but rather, are a product of globalized place-making spurred by consumption.

In the realm of specialty food products from South America, many other examples are seen beyond yerba mate, including açai and quinoa (Keiles 2017; Romero and Shahriari 2011). Beyond food, one of the most striking examples of obscured global supply chains is the smartphone. Does the average Global North consumer understand the resources that go into their smartphone? Are they aware of the conditions under which necessary conflict minerals are mined and integrated into the global economy? Do they understand that physical pieces of Africa are sitting in their pockets and on their desks? Are we, as consumers, unknowingly complicit in any of the unjust or questionable social and environmental processes that are part of the system that allows so many of us to have cell phones? How might corporations benefit from consumer ignorance? If consumers knew, how might that impact the meaning of the product and their consumption of it? Further, what consequences would be spurred by a consumer market more aware of the realities of their consumption?

All these are questions that we as consumers may consider. Put succinctly, this text seeks to answer the following question: How does the pressure to minimize costs in the yerba mate supply chain impact the northeastern region of Argentina, both socially and ecologically? Using a theoretical framework based on the work of geographer Robert Sack, this text will examine the way in which the yerba mate industry defines the product, contrasting that narrative to the social and environmental realities of its production. The narrative paints yerba mate as a product with a long history filled with mutually beneficial relationships between humans and nature. However, with the effort of yerba mate industries to make as much profit as possible, those with the most influence over pricing (the industrial sector) seek to push the price of *hoja verde* (green leaf) down. This forces the actual social and environmental costs down the production chain to those who can, more often than not, afford them least and suffer them most—smallholders. To adapt, I argue that these marginalized smallholders often resort to socially and ecologically destructive practices in hopes of just barely eking out a living. This represents an externalization of both the social and environmental costs of production. The existence of political-economic choices is nothing new to yerba mate. By utilizing an emerging and a rather unconventional commodity as a case study, the effects of the evolving geographies of consumption on Argentina's Atlantic Rainforest region and industry's definition of what yerba mate "is" will be assessed through a political–ecological lens imbued by the work of Sack.

In the case of yerba mate, the social and environmental consequences I focus on include poverty among primary producers, child labor, occupational hazards from herbicide use, and soil erosion that is exacerbated by herbicide use. Consumers often do not directly feel these consequences and, oftentimes, have no awareness of them. By utilizing Ian Cook's "follow the thing" approach, and therefore activating Sack's surface/depth loop, researchers and consumers alike can unshroud the mysteries of the impacts of consumption, allowing for a more accurate understanding of the "good". In this way, Cook's methods allow us to better, "see through to the real".

I conducted fieldwork in Misiones during Summer 2018 and was based out of the city of Posadas, along the banks of the Río Paraná. I conducted a series of seven in-depth semi-structured interviews with diverse stakeholders representing different parts of the production chain and greater society. These stakeholders included people involved in growing, drying, packaging, and marketing yerba mate. The participants were secured mostly through Patricia Ocampo of *Un Sueño para Misiones* (A Dream for Misiones—an NGO seeking the elimination of child labor in the yerba mate production chain). Her assistance was instrumental in my fieldwork. Though the potential for biases in results can arise when working with an NGO, participants represented diverse viewpoints that often did not completely line up with the positions of the NGO.

This method was utilized as it allowed for much more flexibility than structured interviews would have. Though armed with a set of questions at the beginning of each interview, our conversations often drifted to other issues that were not addressed by my pre-written questions. Interviews were recorded and audio files were stored on an encrypted hard drive. Identities of interviewees were anonymized unless stated otherwise. The purpose of these interviews was to utilize a "boots on the ground" approach to identifying the existence of social and ecological threats posed by yerba mate production. Semi-structured interviews allow for much more flexibility, while still following the basic guidelines of the prepared interview questions (Hay 2005). Additionally, as I had never been to this region of Argentina before my fieldwork, I felt it was important to allow for discussion to go in the direction it was led by interview participants. As yerba mate is not the best-studied agricultural commodity, this information is not always readily available or current.

Those interviewed include the following:

Interviewee #1: Patricia Ocampo, *Un Sueño para Misiones* (anti-child labor NGO founder).

Interviewee #2: Male, A marketing manager.

Interviewee #3: Male, A *secadero.*

Interviewee #4: A family of organic yerba mate producers.

Interviewee #5: Male, Executive Director of an industrial firm.

Interviewee #6: Male, A large producer.

Interviewee #7: Male, An agricultural engineer specializing in yerba mate cultivation and processing.

References

Hay I (ed) (2005) Qualitative research methods in human geography, 2nd edn. Oxford University Press, Oxford

Keiles JL (2017) The superfood gold rush. The New York times
Romero S, Shahriari S (2011) Quinoa's global success creates quandary at home. The New York times

Chapter 2
A Biographical Sketch of Yerba Mate

Abstract This chapter provides an overview of yerba mate's use, biogeography, history, and current proliferation. In outlining the recent attempts at the "new" commodity's global spread, I focus on the marketing strategies of several major firms representing various segments of yerba mate's mature domestic market, as well as its emerging global market. Furthermore, I detail the branding efforts of the *Instituto Nacional de la Yerba Mate* (an Argentine quasi-governmental organization) and how they help construct meaning.

Keywords Yerba mate · Argentina · Botany · Biogeography · Marketing · Globalization

2.1 Consumption and Production

Consumption of mate, the drink made with the processed leaves of the yerba mate (*Ilex paraguariensis*) tree, is concentrated in South America, specifically in Argentina (6.5 kg per capita), Brazil (0.8 kg per capita), Paraguay (2.5 kg per capita), Uruguay (8.64 kg per capita), and Chile (0.4 kg per capita) (Barry 2016; Emiliano 2016; Statista 2018). Though the figure for Brazil appears low, consumption is comparable to that of Argentina and Uruguay in certain provinces, particularly along the Uruguayan, Argentine, and Paraguayan borders. In terms of total consumption, Argentina is first in the world, followed by Brazil and Uruguay. The highest consumption per capita is seen in Uruguay (Barry 2016). Outside South America, the only widespread consumption is seen by the Druze people of the Syrian and Lebanese Levant (Folch 2010). Consumption among this group has its roots in the immigration of the Druze to Argentina in the late nineteenth century and early twentieth century. Many Druze quickly developed a taste for yerba mate after arriving in Argentina, then brought the Argentine custom with them when returning to the Middle East.

In addition to being the largest total consumer, Argentina is also the largest producer of yerba mate, with most of the country's production (90%) located in Misiones province in the far northeast of the country (Rau 2009, 51). The remaining 10% of domestic production comes from neighboring Corrientes province (Rau 2009, 51). Roughly 60% of global yerba mate production occurs in Argentina (BILA 2017).

A. S. Dohrenwend, *Green Gold*, SpringerBriefs in Latin American Studies,
https://doi.org/10.1007/978-3-030-82011-4_2

In total, Argentina has just over 165,000 ha devoted to the cultivation of yerba mate, with 144,000 of these hectares located in Misiones and the remaining hectares in Corrientes (INYM 2016). Of Misiones' departments, Oberá is the clear leader, with over 21,000 ha of yerba mate cultivation within its borders. In recent years, the area of cultivated land dedicated to yerba mate has remained stable (Interviewees 2, 3, 4, 5, 6, & 7, 2018).

The table shown in Fig. 2.1 on the next page shows Argentina's total output of dried yerba mate, separated by domestic consumption and export. A large increase in production for domestic consumption is shown, with a spike in the early-mid 1990s. It is worth noting that this large spike occurred concurrently with a significant change in the structure of the Argentine economy: the introduction of the Argentine convertible peso with a value fixed to the US dollar.

On the 165,000 ha of cultivation are around 30,000 farms operated by around 18,000 producers. Of these, 15–16 thousand are small producers, 3,000 are large producers, and around 150 are extremely large producers (Interviewees 2 & 3, 2018). Thirty percent of the producers are responsible for roughly 70% of production, while the remaining 70% of producers are responsible for the other 30% of production

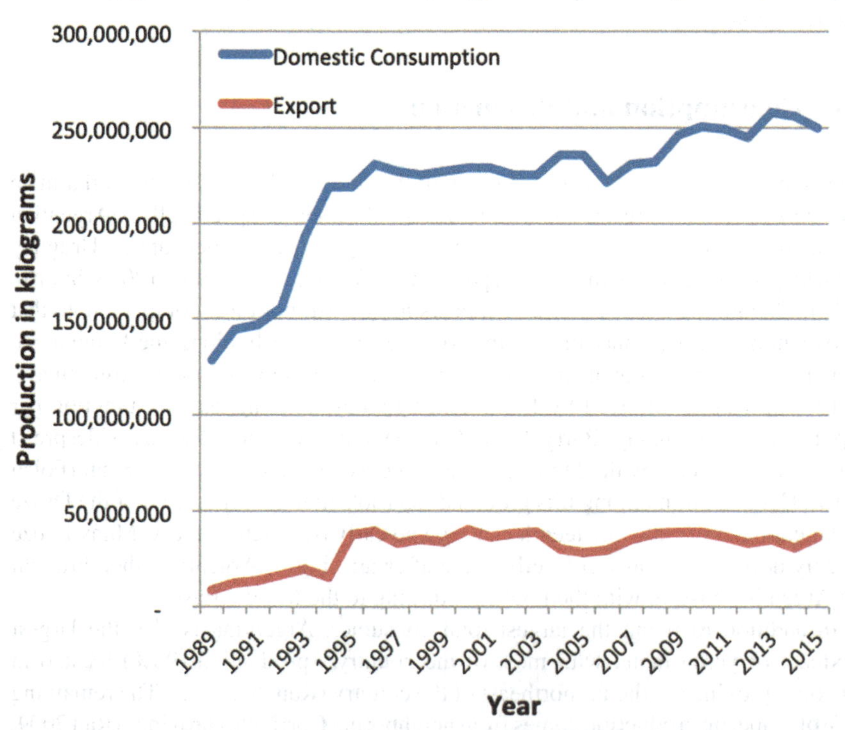

Fig. 2.1 Argentine yerba mate (dry) production—1989–2015 (INYM 2016)

(Interviewees 2 & 3, 2018). Though only 20–30% of total production is by large companies, this figure is much higher when factoring in reliance on contract growing (Interviewee 3, 2018). There are just under 250 *secaderos* (driers) and several dozen large companies in the industrial sector (Interviewee 3, 2018). Of these companies, five hold control over roughly 50% of the market (Interviewee 3, 2018).

Though a bit dated, Rau provides information calculated from INDEC data indicating that 52% of holdings with yerba mate were five hectares or smaller, making up 17% of the total land area devoted to yerba mate (Rau 2009, 52). Holdings between 5.1 and 25 ha made up 42.5% of the total, accounting for about 49% of the total land area (Rau 2009, 52). Holdings greater than 25 ha made up just 5.5% of holdings but contained about 35% of the total area of yerba mate cultivation (Rau 2009, 52).

To prepare mate, the dried and cut leaves are traditionally placed in a dried gourd or similarly shaped vessel. The name "mate" stems from the Quechua word for this vessel, *mati* (Folch 2010). From there, hot water is added and the leaves are briefly steeped. The resulting concoction is then drunk with a straw, referred to as a *bombilla.* In tropical latitudes, much yerba mate consumption takes the form of *tereré*, a cold infusion often mixed with herbs or fruit juices. Yerba mate contains a significant amount of caffeine, along with other compounds, and has been hailed in much biomedical research for its numerous health benefits (Cuelho et al. 2015). Though the spatial extent of large-scale consumption is geographically confined, consumption is generally quite where it does occur.

2.2 Botany and Biogeography of Yerba Mate

Ilex paraguariensis is one member of the *Ilex* genus in the Aquifoliaceae family. The Aquifoliaceae family contains about 400 species of holly that grow on every continent but Antarctica), almost all of which are part of the *Ilex* genus (Porter 1950). There are many Ilex species in the region and it is estimated that roughly one-third have been used in infused beverages (Porter 1950). *Ilex paraguariensis* was classified in 1822 by Augustin Saint-Hilaire, a French botanist. This is where the most common variety gets its common identification as St. Hilaire (Peckolt 1883). He identified three varieties: *obtusifolia, acutifolia,* and *angustifolia* (Porter 1950).

The tree can grow to heights of 20–30 m tall in the absence of human disturbance, however with frequent leaf harvesting, the plant's growth is deliberately stunted to make for a shrub of just 4–6 m in height (Porter 1950). The harvest is completed manually and it is important for producers to ensure their trees do not grow too tall. The species generally live for around 100 years and is dioecious, meaning that each individual is either male or female (Gottlieb 2010; Porter 1950). It is believed that the two different sexes yield final products of different flavors and different caffeine levels (Rakocevic et al. 2012).

The leaves of *Ilex paraguariensis* are perennial, 3–20 cm long, 2–9 cm wide, and an olive-green color. Flowers are generally greenish-white and have four petals and four stamens. These flowers appear in spring (September to November in Argentina)

(Porter 1950). The fruit is an elliptical drupe of about 6.5 mm by 5.5 mm (about the size of a peppercorn). Each fruit contains four seeds and matures in summer (December to March) (Porter 1950). Few seeds are viable (Patiño 1964).

The tree's native range is found between 18 and 25° S latitude in the areas drained by the Paraná, Paraguay, and Uruguay Rivers (Porter 1950). According to Patiño (1964), the best area for growth is in the northeast part of Paraguay on the banks of the Paraguay River. An adaptation of Patiño's 1964 range map by the author is seen in Fig. 2.2. The tree prefers sloping areas with highly weathered, red, acidic, low-fertility Ultisols. This makes cultivated areas especially susceptible to erosion and nutrient degradation (Piccolo et al. 2008). The area's climate features an annual average temperature of 21 °C and over 2000 mm of rainfall each year (SMN 2019). With respect to seasonal variation, winter average temperatures hover around 16 °C while summer average temperatures are around 25 °C (SMN 2019). The region does not experience a dry season; however, the precipitation reaches its peak in summer and autumn (SMN 2019).

2.3 History of Yerba Mate

This narrow growing range along the Paraná-Paraguay River system was home to the Guaraní people (Rosin 2004). During pre-Hispanic times, the Guaraní harvested leaves from the wild stands of yerba mate (Jamieson 2001). This wild harvesting was the predominant mode of the collection as recently as the late 1890s (Jamieson 2001). To consume the leaves for their known medicinal and stimulant effects, the Guaraní simply chewed them, similarly to the use of coca leaves, or drank a concoction created by steeping the leaves and stems in hot water (Girola 1915). The primary medicinal use was to help establish contact with spirits through the work of shamans (Cooper 1949).

Though only grown in the Guaraní area, traces of yerba mate have been discovered in Inca territories to the northwest (Folch 2010). Other tribes known to have consumed yerba mate include the Charrua (largely inhabited modern-day Uruguay), Caingang (Southern Brazil), and the Chamacoco (modern-day northern Paraguay) (Cooper 1949). This suggests that a trade system must have been established to move yerba mate between the Guaraní territories and lands of other tribes as far off as the Andean highlands (Garavaglia 1983).

The Spaniards encountered and developed a taste for yerba mate almost immediately upon their 1537 arrival to the region (Folch 2010). Building off the legends surrounding the plant, the Guaraní used yerba mate as a "sign of hospitality" towards the arriving conquistadors (Oberti 1979).

The Spanish encounter with yerba mate provides an early example of the narrative building by outsider Spaniards to stimulate yerba mate's global commodification. According to colonial lore, the Spaniards first encountered yerba mate when Paraguay's *Criollo* (Latin American born of full or almost full Spanish descent) governor "searched the looted bags of Indigenous Guaraní defeated in a military

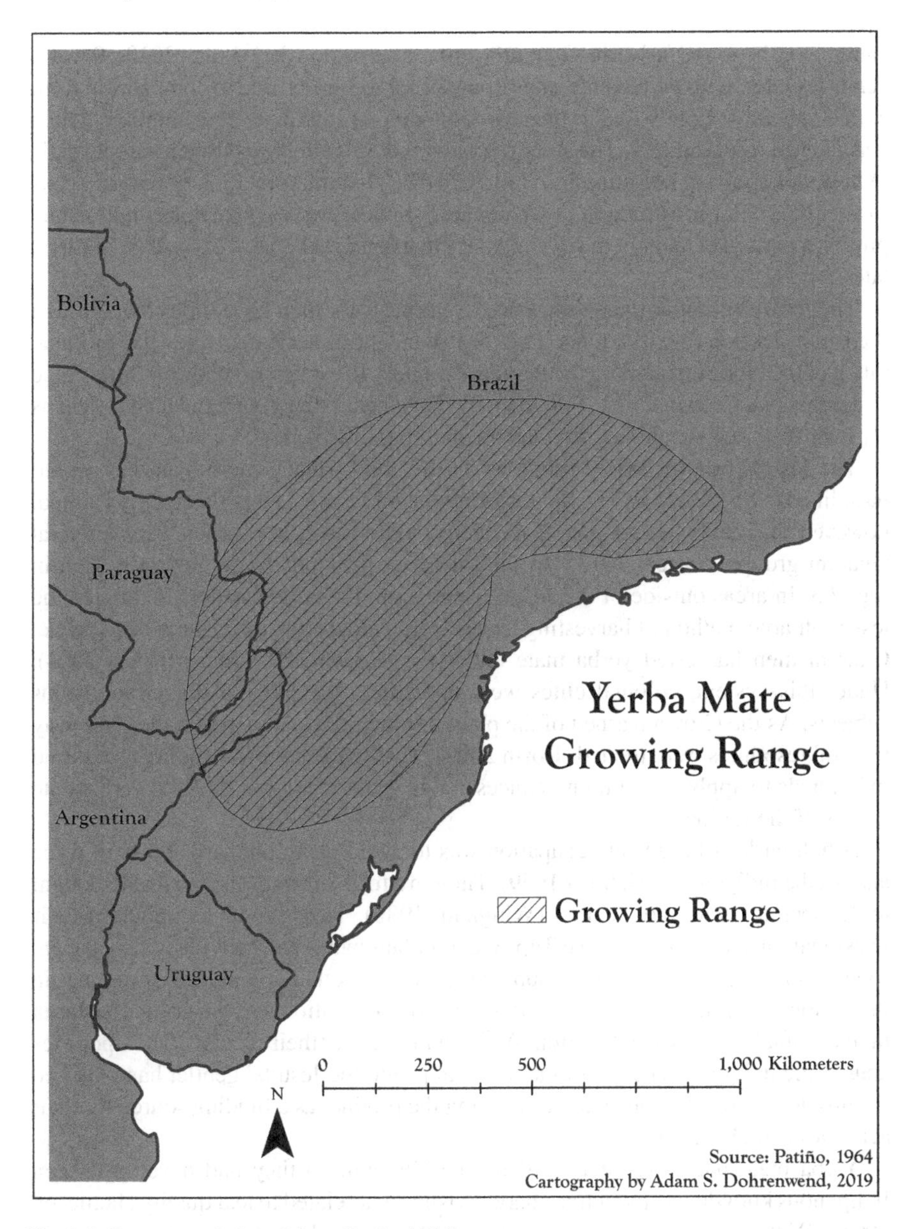

Fig. 2.2 Range map (adapted from Patiño 1964)

campaign," he found *ka'a* (the Guaraní word for yerba powder) (Folch 2010). Several centuries later, with yerba mate sales floundering in Europe due to competition from coffee and tea—largely due to the shared mode of yerba mate consumption, which was viewed as unsanitary. The story of colonial domination was used as propaganda in hopes of spurring consumption (Folch 2010). This interpretation of history seeks to alter the social construction of what mate is in order to heighten consumption and profits, a pattern evident in today's industry marketing scheme. This will be outlined later.

The Jesuits serendipitously came across the basis for their economic clout through their introduction to yerba mate. Lacking abundant mineral resources like Bolivia, much of the land surrounding the Paraná-Paraguay River system had very little value to the Spanish Crown at first (Folch 2010). This meant that giving the land to Jesuits for their missions would not threaten income to the Crown.

The Jesuits established 32 missions on this land—many more than they would have liked (Sarreal 2013). The Spanish Crown and Jesuit leadership quickly learned, however, that having fewer, larger missions would result in conflicts between rival Guaraní groups (Sarreal 2013). A map of Jesuit missions in the region is seen in Fig. 2.3. In areas outside of the Jesuit occupation, the conquistadors continued the less profitable method of harvesting leaves from wild tree stands. Under this regime, Guaraní men harvested yerba mate under the *encomienda* system (Rosin 2004). Under this system, colonial elites were entitled to the labor of their Indigenous subjects. As the Crown learned of the plant's value, it began to treat it the same way mineral resources were treated (Rosin 2004). The Spanish crown's policies required individuals to apply for a mining concession in order to exploit the wild yerba mate groves of the region.

A hallmark of the Jesuit occupation was to avoid using outward shows of force against the Indigenous (Métraux 1949). Though a form of conquest, the Jesuits sought to "cooperate" with the Guaraní (Whigham 1986). They conducted much of their missionary activity in the native Tupi-Guaraní languages and provided fairly decent food rations and clothing to their subjects, relative to other colonial labor regimes in the region (Métraux 1949). This spirit of stated cooperation was the critical factor in their eventual success in cultivation. While claiming that their secular contemporaries abused the Indigenous peoples as a means of profit, the Jesuits' gentler hand nudged the missions towards mate extraction, using the product as a funding source for their activities (Folch 2010).

Yerba mate had long been used by the Guaraní, so they had developed keen Indigenous knowledge about tree characteristics that related to leaf quality (Jamieson 2001). With forced Guaraní labor, the Jesuits transplanted trees from the wild stands to their mission sites and started the evolution towards plantation agriculture (Whigham 1986). As pioneers of yerba mate cultivation, largely by exploitation of the Guarani's Indigenous knowledge, the Jesuits utilized many tricks to assist with growth (Jamieson 2001). Some included only harvesting seeds that have aged until purple in color, soaking them in water to eliminate the non-viable seeds, and then letting the seeds pass through the digestive systems of birds, commonly toucans (Folch 2010; Patiño 1964).

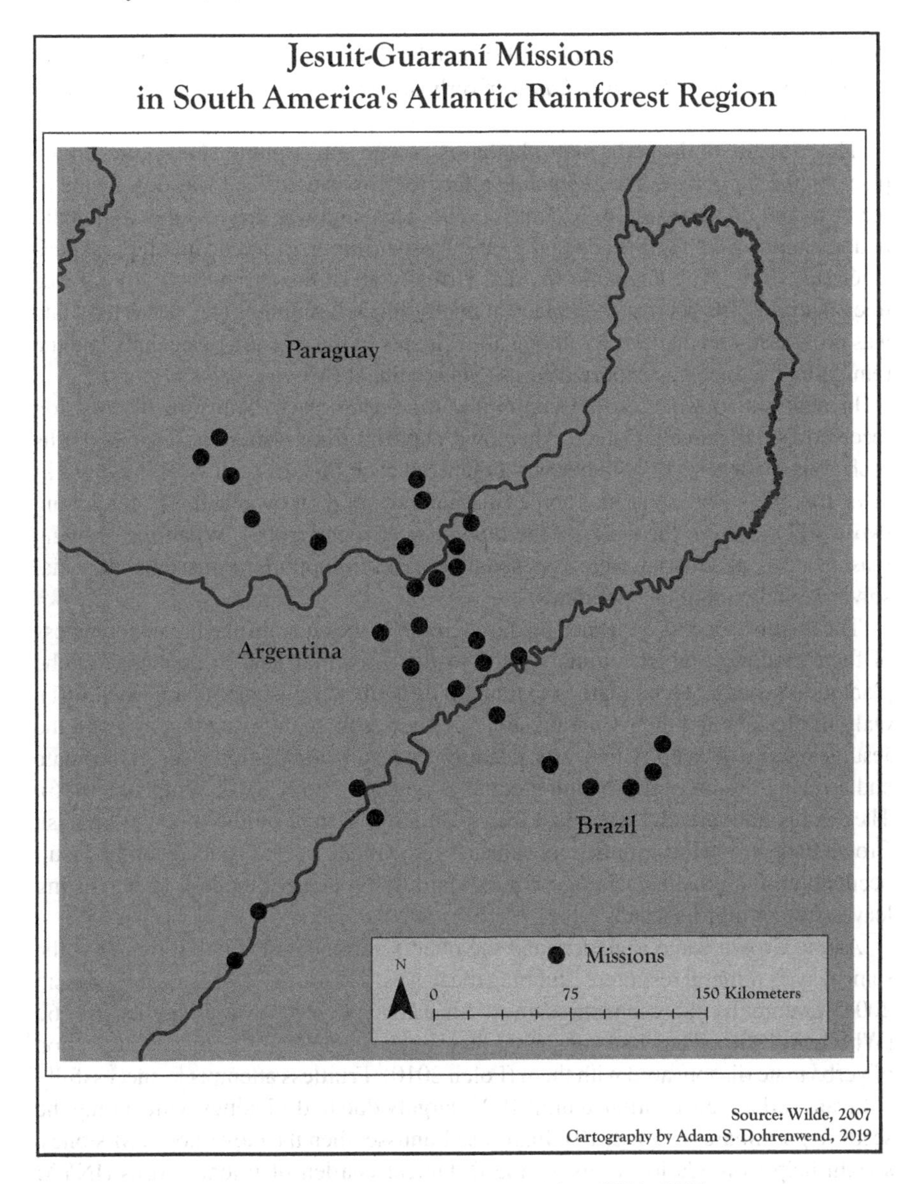

Fig. 2.3 Jesuit missions in the yerba mate region (adapted from Wilde 2007)

The Jesuits began trading large quantities of yerba mate across the Viceroyalty of the Río de la Plata and the Viceroyalty of Peru once cultivation practices were firmly established (Crocitti 2002). The product was not viewed by the Jesuits as a product for export overseas and strong followings for Jesuit mate were developed in cities like Tucumán, Potosí, and Lima (Crocitti 2002; Patiño 1964). Mate sold in Lima for four to five times higher than in Buenos Aires (Crocitti 2002). The shipping of yerba mate to these other locations was undertaken by the Guaraní as well. Bands of them would often move by *balsas* or *itapas* (riverboats) with the finished product (Crocitti 2002).

The structure of the yerba mate plantation system was communal in nature, taking place on the *tupambae*. The *tupambae* refers to "God's acre" and was designated as the best land on the mission (Métraux 1949). This land was cultivated communally under a land tenure regime adapted by the Jesuits from traditional Guaraní practices (Métraux 1949). The Guaraní provided virtually all of the agricultural labor. Once trees matured, the harvest took place annually for 3–4 months. The harvested mate was processed and distributed among all residents of the mission twice daily and the remaining product was exported across the continent (Métraux 1949).

In addition to work on the communal *tupambae*, each family in the mission received a small parcel of land for their own staple crop cultivation for several months each year (Crocitti 2002). Labor was organized such that, of each week's six workdays, four days were set aside for the Guaraní to work their own land. The remaining two workdays were for work on the *tupambae* (Crocitti 2002). When the growing season ended, all Guaraní were assigned other tasks to support the missions' financial solvency and continued operations.

The Jesuits' success depended on the coercion of the Guaraní and the secretiveness of their practices, which ultimately led to their downfall and the collapse of the plantation system. Yerba mate is extremely difficult to transplant (Jamieson 2001). Only the Jesuits and their Guaraní subjects knew how to cultivate the crop and the Jesuits sought to keep it this way, leading to Jesuit domination over yerba mate and drawing the ire of surrounding colonial powers (Folch 2010; Whigham 1986). The Jesuits had already developed their plantation system by the time the Spanish Crown understood the commercial value of yerba mate. In areas outside of the Jesuit occupation, the conquistadors continued the less productive method of harvesting leaves from wild tree stands.

As the Crown came to understand the plant's commercial value, it treated it the same way as mineral resources, dubbing the leaves "*oro verde*" ("green gold") (Rosin 2004). Eventually, these tensions came to a head and the Jesuits were expelled in 1767 (Whigham 1986). They had not shared their trade secrets and significant cultivation of yerba mate disappeared with them (Folch 2010). Fruitless attempts to successfully cultivate yerba mate continued until 1903, largely due to difficulties with getting the seeds to germinate (Folch 2010). Juan José Lanusse, then the Governor of Misiones, sought help from Carlos Thais, of the Botanical Garden of Buenos Aires (INYM 2014). Thais discovered that germination could be achieved if the seeds were treated with hot water, and this technique was publicized by Governor Lanusse (INYM 2014). The first commercial plantation was founded in 1903 in San Ignacio, along the Río Paraná.

Consumption had skyrocketed due to prior abundance under the Jesuit plantation regime, and royal authorities had to find a new way to supply the region's literal addiction to yerba mate. The Guaraní were nothing more than slaves to the royal administrators, a slight change in attitude from the Jesuits (Whigham 1986). This effectively led to a refusal to share the secrets of the Jesuits with their new rulers.

Since cultivation was not possible without the Jesuit and Guaraní knowledge, royal authorities reverted to the exploitation of wild stands (Jamieson 2001). This exploitation was much more intense in order to fulfill higher demand, thus, it can be argued that the political-economic decision to expel the Jesuits led directly to the environmental degradation of the forests and the much harsher treatment of Indigenous groups. Intense pruning methods removed much of the tree's branches and limited regeneration (Rosin 2004). The severe degradation of the forests meant that their yields were not capable of fulfilling domestic demand.

After the region's independence movements of the early 1800s, newly formed Argentina had millions of mate drinkers but no domestic source of production, at least until the War of the Triple Alliance (Rosin 2004). The war, fought between 1864 and 1870, was based on territorial and resource concerns in the Southern Cone and involved Argentina, Brazil, Paraguay, and Uruguay. Argentina had a direct interest in gaining control over a sliver of land in yerba mate's narrow growing region, as it would be able to domestically fulfill the country's demand for yerba mate. Without that land, Argentina would be at the mercy of hostile neighbors to fulfill their internal demands (Rosin 2004).

Tensions related to Argentine and Brazilian influences on Uruguayan democracy instigated a war, in which yerba mate played an indirect role. In response to Brazil's meddling in Uruguay, Paraguay invaded, crossing through Misiones, an area disputed by Argentina. In response, Argentina, Brazil, and Uruguay (the Triple Alliance) joined together and decimated the Paraguayan forces (Rosin 2004). The war's toll on Paraguay was severe both in terms of loss of life and of territorial extent; two-thirds of Paraguay's male population was killed and the Misiones area was ceded to Argentina (Rosin 2004). It is also worth noting that this war and Argentina's participation, largely over territorial influence and control, were particularly hard on the region's Indigenous peoples. With Misiones under its national jurisdiction, Argentina finally had territory within yerba mate's natural growing range and could secure self-reliance in production.

Once cultivation returned, Argentina quickly moved to consolidate production in the mate-growing zone that was acquired through the War of the Triple Alliance, the new Argentine provinces of Misiones and Corrientes (Rosin 2004). Argentina, the largest consumer, choked off imports from Paraguay and Brazil and developed significant government support programs to foster production increases (Rosin 2004). This led to the country becoming the largest producer as well, with 90% of this domestic production occurring in the narrow Misiones province (Rau 2009).

Starting in 1910, the Argentine government began directly subsidizing plantations and later established a government research institute for yerba production, in hopes of fulfilling the goal of self-sufficiency (Rosin 2004). This institute, called the Instituto Nacional de Yerba Mate (INYM) in its current form, is located in Posadas, the capital city of Misiones located on the banks of the Río Paraná. INYM is "a non-state entity of public law with jurisdiction throughout Argentina" that was organized in 2002 by a vote of the federal legislature to represent the interests of Argentine yerba mate growers and processors, as well as to stabilize prices for the purpose of protecting farmers (Interviewees 2 & 3, 2018; Ley Argentina 25.564, 2002). Its governing board consists of 12 members: one appointed by the President of Argentina, one each appointed by the Governors of Misiones and Corrientes (the provinces where mate is grown), two from the industrial sector, three from the primary grower associations, two from the agribusiness sector, one from the dryers, and one to represent rural workers (Ley Argentina 25.564, 2002). All yerba mate produced and sold in Argentina is under the auspices of INYM, with each package bearing a stamp and serial number.

2.4 Production Process

Today, yerba mate is largely cultivated through plantation agriculture. The plantation landscape is reminiscent of an apple orchard. The process begins with submerging the seeds in water for around nine days. After, the seeds are dried for one to two months before they are planted. After roughly a year of growth, the developing *plantitas* (juvenile plants) are sold to *yerbateros* (small producers) and are then planted in rows with 1.5–3 m of space in between each individual. Typically, one hectare can accommodate 1,650 plants. Plants enter production between two and four years after planting when they undergo the first trimming (Interviewee 3, 2018).

The harvest season generally occurs between April and September (Argentina's fall and winter), as yields are higher during this time. From April to September, this value hovers between 34 and 37%, while harvest during the spring and summer months yields between 29 and 32% (Montechiesi 2016). This means that the period from April to September provides the most efficient harvest, as it takes less *hoja verde* during this time to produce a kilogram of *hoja seca.* These data show that it takes between 2.7 and 3.5 kg of *hoja verde* (green leaf) to produce one kilogram of *hoja seca* (dried plant material) (Figs. 2.4 and 2.5).

With most of the production occurring during the more favorable months, the average is 2.857 kg (Montechiesi 2016). These percentage values are calculated by dividing the weight of the *hoja seca* by the weight of the *hoja verde* and then multiplying the quotient by 100 (Montechiesi 2016). Harvest frequency varies by the individual producer. The three most common harvest regimes are every year, every year and a half, and every other year. Harvesting every other year allows for a more complete regeneration of the tree's leaves and up to 30% higher yield over the yearly harvest (Montechiesi 2016). Though harvest every other year generates sustained revenue for farmers over time, many feel economic pressure to harvest as often as possible to get by (Interviewees 3, 4 & 7, 2018). Cultivated land use in the area is diverse, with many farmers producing more than one crop. Some common cultivars in the area include citrus species, eucalyptus, pine, tobacco, tea, and tung (Interviewees 3, 4, & 7, 2018).

Harvest is generally manual, often assisted by scissors or small saws. To spur radial growth, trees are pruned frequently from a young age in order to stunt their vertical growth, creating what was referred to during an interview as desirable *"árboles*

Fig. 2.4 A newly planted *plantita*, Oberá (photograph by the author)

gordos" ("fat trees") (Interviewee 4, 2018). Manual harvest becomes difficult if trees are allowed to grow too tall.

From the *yerbales* (yerba mate fields), the *hoja verde* is transported to a *secadero* to be dried. The yerba arrives at the drying facility within 24 h of harvest where direct fire is applied (*zapecado*). After about 90 s, the leaves lose 30-40% of their water content and their internal enzymatic and biological processes are mostly halted. The leaves then go through the primary drying process in which hot air from furnaces is applied as they pass repeatedly along a conveyor belt (Montechiesi 2016). The furnaces were historically powered by wood from the surrounding forests, a practice that shifted in recent years to the use of pellets produced as a byproduct of the

Fig. 2.5 An organic yerba mate farm, Oberá (photograph by the author)

region's paper and lumber industries (Interviewees 3 & 7, 2018). After drying, the plant material is ground up and put into 50-kg burlap sacks for aging. The yerba mate develops flavor and color during this period. Historically, this process took six months to as much as two years, but the development of accelerated methods shortens the period to under 2 months (Montechiesi 2016).

2.5 Marketing, Global Proliferation and INYM

The long history of yerba mate's stunted growth on a global scale started with its attempted introduction in Europe. Its prospects were limited by two main factors. First, other stimulant beverages (coffee and tea) had already been introduced. Second, the European market found the shared nature of the beverage unappealing for sanitary reasons. In the Southern Cone, consumption involves the shared use of a single *bombilla* by the entire group, leading one to confess, "it was unpleasant to put into my mouth the unclean tip of the pipe-like stem through which the mate drink was sucked" (Albes 1916, 11). This illustrates the lack of comfort Europeans still in Europe had with the traditional Indigenous method of consumption, methods that persist today throughout the Southern Cone.

In this way, yerba mate consumption draws a distinct dissimilarity with the consumption of other stimulant beverages. Coffee and tea are consumed individually, while the social significance of consuming yerba mate in a group is seen by some as unsanitary. It is rather ironic that the New World's disease vectors (Europeans) responsible for the epidemics that decimated Indigenous populations during the early colonial area were the ones simultaneously concerned about catching a virus or spreading bacteria from Indigenous practices of consumption.

The Argentine consumer market is saturated in terms of yerba mate consumption and the industry has begun seeking new markets for its product in the industrialized world. At the center of yerba mate's global marketing campaign is INYM and its trade publications. From here, pro-mate rhetoric is proliferated globally through their English-language website and through the marketing of international companies.

When perusing INYM's website, its trade publications, and the marketing of yerba mate companies directed at their desired markets, several patterns are evident. First is the picture that yerba mate is consumed around the world. In INYM's 2015 annual report, a list of trade shows with the association's presence is provided. Cities included in the list of junkets are Dubai, Seoul, Ho Chi Minh City, New York, Lima, Santiago, and Paris (INYM 2015). In each of these locations, INYM's representatives pledged support to help commercialization and preached that their product was 100% natural, *"desde la planta hasta ser envasado"* (from plant to packaging), had numerous nutritional benefits deriving from high concentrations of antioxidants, vitamins, and minerals (INYM 2015). The product is also versatile, allowing for numerous methods of consumption including cold preparations and being blended with fruit (INYM 2015). The industry's marketing campaign can be summed up as the following: consuming yerba mate is good for mind, body, and spirit—as well as for those in the growing area. The globalizing nature of the Argentina yerba mate industry is shown by export figures from INDEC found in the statistical appendix.

Is it true that, in the words of INYM, "Mate means friendship, sharing, and connecting with yourself and others" (INYM 2017)? Is yerba mate the quintessential natural product? Is a blogger's intuition correct when they describe their experience trying a handful of yerba mate products by writing, "...consuming these products instills a feeling of being Mother Nature's favorite child" (Smith 2010)? It certainly seems that way on the websites of Guayakí, and Taragüi, two of the largest brands in global yerba mate distribution. The marketing employed implies that consuming yerba mate improves yourself and the environment, as well as provides a decent living for those involved in the supply chain.

Guayakí claims that its mission is "to steward and restore 200,000 acres of South American Atlantic Rainforest and create over 1000 living wage jobs by 2020 by leveraging our Market Driven Restoration Model" (Guayakí 2017). They also claim that consuming their products may actually reduce one's carbon footprint, that their work makes the environment and farmers thrive, and that their product "provides a joyful experience that helps [consumers] discover and embrace the cultural traditions and spirit of sharing that is at the core of the healthy yerba mate lifestyle" (Guayakí 2017). Guayakí's website is completely in English and the company is based in California (Guayakí 2017). While the US beverage companies do sell traditional yerba mate sold loose, their main focus is a large array of bottled yerba mate drinks, now even including energy shots (Guayakí 2017).

Taragüi's website also has its share of content to increase its global appeal. It asserts that "Mate is unique and everyone should try it," and features photographs of Pope Francis drinking yerba mate. The site even has a "Which mate suits your personality?" Buzzfeed-style quiz (Taragüi 2017). The company is based in Corrientes, Argentina and its website provides full versions in English, Spanish, Korean, Italian, French, Dutch, Polish, and Swedish (Taragüi 2017).

While the validity of these companies' bold claims deserves examination, this work simply seeks to compare this discourse with some of the social and environmental realities experienced in Misiones. These include the effects of monocultures on soil degradation and erosion, as well as the province's sustained poverty.

References

Albes E (1916) Yerba mate: the tea of South America. In: Bulletin of the Pan American Union. Pan American Union, Washington D.C.

Barry M (2016) Chicory, mate, and beyond: opportunities in traditional plant-based hot drinks. In: Strategy reports. Euromonitor, London, U.K., pp 1–41

BILA (2016) List of goods produced by child labor or forced labor. Untited States Department of Labor.,Washington D.C.

Cooper JM (1949) Stimulants and narcotics. In: Steward JH (ed) Handbook of South American Indians. United States Government Printing Office, Washington DC, pp 525–558

Crea El Instituto Nacional De La Yerba Mate. Ley N° 25.564. 14 March 2002.

Crocitti JJ (2002) The internal economic organization of the jesuit missions among the guaraní. Int Soc Sci Rev 77(1/2):3–16

Cuelho CHF, de França Bonilha I, do Canto GS, Manfron MP (2015) Recent advances in the bioactive properties of yerba mate. Revista Cubana de Farmacia 49(2):375–383

Emiliano L (2016) Aumenta El Consumo Per Cápita De Yerba Mate En Uruguay Pero Cae El Consumo En Chile. Agroindustrias del NEA

Folch C (2010) Stimulating consumption: yerba mate myths, markets, and meanings from conquest to present. Comp Stud Soc Hist 52(1):6–36

Garavaglia JC (1983) Mercado Interno Y Economia Colonial. Editorial Grijalbo, Mexico, D.F.

Girola CD (1915) El Cultivo De La Yerba Mate (Ilex Paraguariensis). Buenos Aires

Gottlieb AM (2010) Genomic screening in dioecious "Yerba Mate" tree (Ilex Paraguariensis A. St. Hill., Aquifoliaceae) through representational difference analysis. Genetica 138(6):567–578

Guayakí (2017) Guayakí Sustainable Rainforest Products, Inc. http://guayaki.com/

Interviewee 2 (Marketing Manager) (2018) Interview by A.S. Dohrenwend. Semi-structured interview. Posadas, July 2018.

Interviewee 3 (*Secadero*) (2018) Interview by A.S. Dohrenwend. Semi-structured interview. Posadas, July 2018.

Interviewee 4 (Family of organic producers) (2018) Interview by A.S. Dohrenwend. Semi-structured interview. Oberá, July 2018.

Interviewee 5 (Executive Director of large industrial firm) (2018) Interview by A.S. Dohrenwend. Semi-structured interview. Posadas, July 2018.

Interviewee 6 (Large producer) (2018) Interview by A.S. Dohrenwend. Semi-structured interview. Posadas, July 2018.

Interviewee 7 (Agricultural engineer) (2018) Interview by A.S. Dohrenwend. Semi-structured interview. Posadas, July 2018.
INYM (2014) Bien nuestro. Instituto Nacional de la Yerba Mate, Posadas
INYM (2015) Anuario. Instituto Nacional de la Yerba Mate, Posadas
INYM (2016) Superficie Cultivada Por Departamentos. Instituto Nacional de la Yerba Mate, Posadas
INYM (2017) ¿What is yerba mate? INYM. http://yerbamateargentina.org.ar/en/yerba-mate/que-es-la-yerba-mate/
Jamieson RW (2001) The essence of commodification: caffeine dependencies in the early modern world. J Soc Hist 35(2):269–294
Métraux A (1949) Jesuit missions in South America. In: Steward JH (ed) Handbook of South American Indians. United States Government Printing Office, Washington DC, pp 645–653
Montechiesi R (2016) Yerba Mate De Ayer, De Hoy Y De Siempre. Posadas
Oberti F (1979) Historia Y Folklore Del Mate. La Imprenta del Congreso de la Nación, Buenos Aires
Patiño VM (1964) Plantas Cultivadas Y Animales Domesticos En América Equinoccial, vols 3 and 6. Imprenta Departamental, Fibras, medicinas, miscelaneas, Cali, Colombia
Peckolt T (1883) Mate or paraguay tea. Am J Pharm 55(2):570–575
Piccolo GA, Andriulo AE, Mary B (2008) Changes in soil organic matter under different land management in Misiones Province (Argentina). Scientia Agricola 65(3):290–297
Porter RH (1950) Maté—South American or paraguay tea. Econ Bot 4(1):37–51
Rakocevic M, Janssens M, Schere R (2012) Light responses and gender issues in the domestication process of yerba-mate, a subtropical evergreen. In: Bezerra AD, Ferreira Bezerra TS (eds) Evergreens: types, ecology and conservation. Nova Science Publishers, Hauppauge, NY
Rau V (2009) La Yerba Mate En Misiones (Argentina). Estructura Y Significados De Una Producción Localizada. Agroalimentaria 15(28):49–58
Rosin CJ (2004) The political ecology of Mercosur/L: local knowledge and responses to a competitive market. University of Wisconsin-Madison
Sarreal J (2013) Revisiting cultivated agriculture, animal husbandry, and daily life on the Guaraní Missions. Ethnohistory 60(1):101–124
Smith D (2010) Guayaki yerba mate energy drink review. Caffeine Informer
SMN (2019) Clima En La Argentina: Oberá. Servicio Meteorológico Nacional.
Statista (2018) Annual per capita consumption of yerba mate in selected countries in Latin America as of July 2018 (in Kilograms). Statista, Hamburg
Taragüi (2017) Establecimiento Las Marías. https://www.taragui.com/en/
Whigham TL (1986) The politics of river commerce in the Upper Plata, 1780–1865. Stanford University.
Wilde G (2007) Toward a political anthropology of mission sound: paraguay in the 17th and 18th centuries. Music Politics 1(2):1–29

Chapter 3
"Get[ting] Behind the Veil:" A "Sackian" Political-Ecological Approach

Abstract This chapter illustrates the volume's theoretical framework for understanding and assessing the marketing strategies presented in Chap. 2. Here, political ecology is discussed, and the approach is imbued with key contributions by geographers Robert Sack and Ian Cook, as well as sociologist Michael Carolan. They argue that the approach laid out can help build a more accurate understanding of distant place-making spurred by consumption, or, in Harvey's words, to "get behind the veil" (Harvey in Ann Assoc Am Geogr 80:418–434, 1990).

Keywords Political ecology · Robert Sack · Place-making · Advertising · Food systems

3.1 Political Ecology

Political ecology is a framework that politicizes and analyzes human-environmental interactions (Rosin 2004). The lens has its roots in cultural ecology, a theoretical approach pioneered by anthropologist Julian Steward (Gunn 1980; Robbins 2012). Cultural ecology is defined by Castree et al. (2013) as "the study of the relations between humans and their environment, paying particular attention to processes of adaptation through cultural means." By the 1980s, critics of cultural ecology in geography contended that the framework ignored the underlying political-economic structures and pressures of a globalizing world (Robbins 2012). In response, Blaikie (1985) published the seminal work on political ecology in geography: *The Political Economy of Soil Erosion in Developing Countries* (Rosin 2004; Blaikie 1985). Two years later, Blaikie and Brookfield (1987) published a related edited volume: *Land Degradation and Society.* Their work focused on erosion in Africa and Asia, setting the stage for the framework's general focus on the Developing World (Zimmerer and Bassett 2003).

Blaikie and Brookfield's ideas sought to tie environmental change to the processes governing the global political economy. They argued that external political factors such as state policy played a significant role in landholders' land-use decisions (Robbins 2012). Their work was filled with complex diagrams and flowcharts to illustrate chains of explanation (Robbins 2012). In constructing these chains of

A. S. Dohrenwend, *Green Gold*, SpringerBriefs in Latin American Studies,
https://doi.org/10.1007/978-3-030-82011-4_3

explanation, Blaikie and Brookfield utilized the human ecology principle of progressive contextualization (Vayda 1983). This principle implores researchers to not view processes in isolation. Rather, they ought to seek out answers about a seemingly never-ending series of related events operating at different scales of human organization (Vayda 1983).

Today, two of the most influential geographers utilizing the political–ecological approach in their work are Karl Zimmerer and Paul Robbins. Both have published widely used texts on the framework, in 2003 and 2004, respectively. Zimmerer states that "Geographical political ecology focuses on socio-natural scaling which occurs in the fusing of biogeophysical processes with broadly social ones" (Zimmerer 2000, 153). He is particularly concerned with the role of scale and focuses on local–global processes, or "glocalization" (Zimmerer 2000). This centrality of scale is critical, as it illustrates the reciprocal, multiscalar interplay between natural, political, and market forces at global, regional, and national levels with local processes. Zimmerer's sustained interest in biogeophysical processes is particularly apparent in his former advisee Rosin's comprehensive work on yerba mate as a means of analyzing MERCOSUR, a political and trade bloc including Brazil, Argentina, Paraguay, Uruguay, and Venezuela (currently suspended) (Rosin 2004).

Robbins provides a different perspective that has spurred significant criticism of political ecology. Rather than defining the framework in specific terms, Robbins's approach defines political ecology as a community of practice and as "something that people do" (Robbins 2012, 4). Robbins's writings focus much more on the political than on the ecological. This is a common trait among other prominent political ecologists, like Michael Watts (Perreault and Watts 2011). This trend has led some to disparage political ecology as "politics without ecology" (Zimmerer and Bassett 2003, 103). Walker examined this claim and, while finding it rather dubious, acknowledges that it exemplifies the overall trend in the field (Walker 2005). He argues that though many political ecologists pay much less attention to ecological principles than during the years of Blaikie and Brookfield, claims of "politics without ecology" are "premature" (Walker 2005, 78).

Vayda and Walters (1999) echo the sentiment that political ecology has become "politics without ecology." They argue that the founding of political ecology was a response to the lack of attention to "political dimensions" by human ecology when examining interactions with the environment—so-called "ecology without politics" (Vayda and Walters 1999). The most troubling aspect of political ecology, according to the authors, is that it assumes that the wider political-economic system is "always important" and often the single most important factor, leading researchers to discount ecological principles and interactions (Vayda and Walters 1999). Using Vayda's own foundational principle of progressive contextualization should lead researchers to acknowledge all potential realms involved in change across scales. Outside sociopolitical factors may be paramount in one context, but this should not be set as a general rule as it has been in much political ecological research.

Though there are different perspectives on what political ecology is, an effective model inclusive of these different viewpoints is captured in the model of the Kite, as proposed by Campbell and Olson (1991). They propose a diamond (the points of the

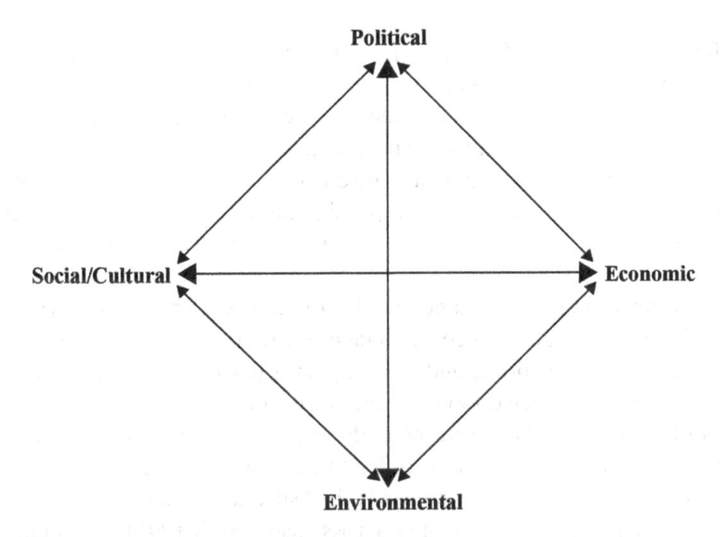

Fig. 3.1 Visualizations of the Kite (adapted from Campbell and Olson 1991)

Kite) with different categorical variables at each of the four points: economic, social-cultural, political, and environmental (Campbell and Olson 1991). All four points are connected to one another through the Kite's frame, consisting of scale, space, power, and time (Campbell and Olson 1991). This framework visually represents the basis of the political–ecological lens: the interactions between humans and their environment viewed in a politicized manner that recognizes temporal and scalar implications. Figure 3.1 shows this visualization (Campbell and Olson 1991).

3.2 Robert Sack and the Construction of Place

Applying the ideas of geographer Robert Sack is a relatively straightforward goal when constructing a political–ecological lens to examine yerba mate. To spur consumption, a context must be constructed for consumers to decide to buy mate instead of something else (coffee, tea, juice drink, etc.). As Sack argues, the symbols and representations of the commodity become the commodity. Though people purchase the material object, it can be argued that the real purchase is what the product means for them. People purchase a product not for the material value, but rather for both what it can make them and where it can place them. Sack addresses these ideas throughout his books *Consumer's World* (1992) and *Homo Geographicus* (1997).

Sack's conception of place as a force is detailed in *Homo Geographicus*. His premise begins with a key assertion in the first sentence: "We humans are geographical beings transforming the earth and making it into a home, and that transformed world affects who we are" (Sack 1997). People are, by nature, geographical actors, and play a powerful role in the creation of place and its forces.

Sack's writing focuses on place, which weaves three separate realms into one: meaning, nature, and social relations. These realms may then take on varying degrees of importance in the construction of a given place. For example, Sack describes national parks as being places that emphasize nature, universities as places that emphasize meaning, and prisons as places that emphasize social relations (Sack 1997).

Places are active and act as a force. For this to occur, each realm and its associated loop, which when activated, forces a change in the other loops and the particular mix of nature, social relations, and meaning in a place at a particular time. The weave of these interconnected loops is what forms the backbone of a place. Within the realm of meaning is the "surface/depth loop," which involves the questioning of superficial qualities of a place in order to get a better understanding of reality (Sack 1997). Within the realm of nature is the "spatial interaction loop," involving the movement and interaction of matter across space (Sack 1997). Within the realm of social relations is the "in/out-of-place loop," which has to do with the rules about what should and should not occur or exist in a place (Sack 1997).

In later work, Sack applies this foundational empirical framework to examine how a place can lead us to an understanding of the good (Sack 2003). Sack introduces two kinds of morality—instrumental and intrinsic—and argues that in order to steer clear of the traps of relativism and absolutism, intrinsic values must take precedence. Without the favoring of intrinsic values, we are left with only instrumental judgments, which can allow for arguing that Auschwitz was a moral place, based solely on the notion that it was very effective at achieving its instrumental goal. The two intrinsic values of a moral place are that it allows for a better understanding of reality (what Sack calls "seeing through to the real" and which heavily relies on the principle of questioning inherent to the surface/depth loop) and that it increases variety and complexity or reality (Sack, 2003). These two criteria often check each other, as both must be met. For example, if a policy is desired by a state to foster or protect biodiversity in an at-risk area, but can only be achieved through authoritarianism, it would not make for a more moral place because a culture of questioning would be stifled. In this case, a better understanding of reality would not be promoted.

The thin places of our hyper-connected global economy can obscure the consequences of our actions. According to Sack (2001), thin places are those that are hyper-specialized, limiting the prospects of developing shared experiences in the construction and maintenance of places. On the other hand, thick places are those in which many different activities occur together in a multi-layered fashion. For example, a pre-modern hunter-gatherer may have a campfire in which political discussions, entertainment, education, and meals all occur (Sack 2001). These thick places are imbued by layers of shared experiences.

Sack argues that the thinning of place is a central characteristic of modernity. Places of production now largely consist of highly specific activities, with each place designed to add value to a given end product. As flows among places become more impersonal and segmented into hyper-specialized places of production to add value, the risk of consumers losing touch with the spatial flows that make their consumption possible is augmented. This ignorance can serve as an unknowing endorsement of unsavory social and environmental impacts of consumption in these distant hyper-specialized places of production. Knowing this, geography can serve as a means of reconstructing these spatial flows (Fig. 3.2).

In *Place, Modernity, and the Consumer's World*, Sack argues that companies construct new realities for their consumers through advertising and influence on popular culture (1992). Advertising and cultural influence motivate the production, movement, and consumption of products across markets. Why consume yerba mate over the countless other products on the market? Advertising supplies the answer.

Products symbolize much more than their direct function as social contexts intertwine to define them. According to Sack, "each ad becomes an idealized picture of how commodities empower us to create contexts" (Sack 1992). In the case of yerba mate, the consumers are lured into purchasing the constructed meaning of mate and how it translates into their own reality, rather than the material itself. Furthermore, consumption can facilitate consumer connection to locations and nature.

Sack provides two particularly relevant examples with a General Motors advertisement for its GMC Jimmy truck model and an advertisement for Royal Doulton

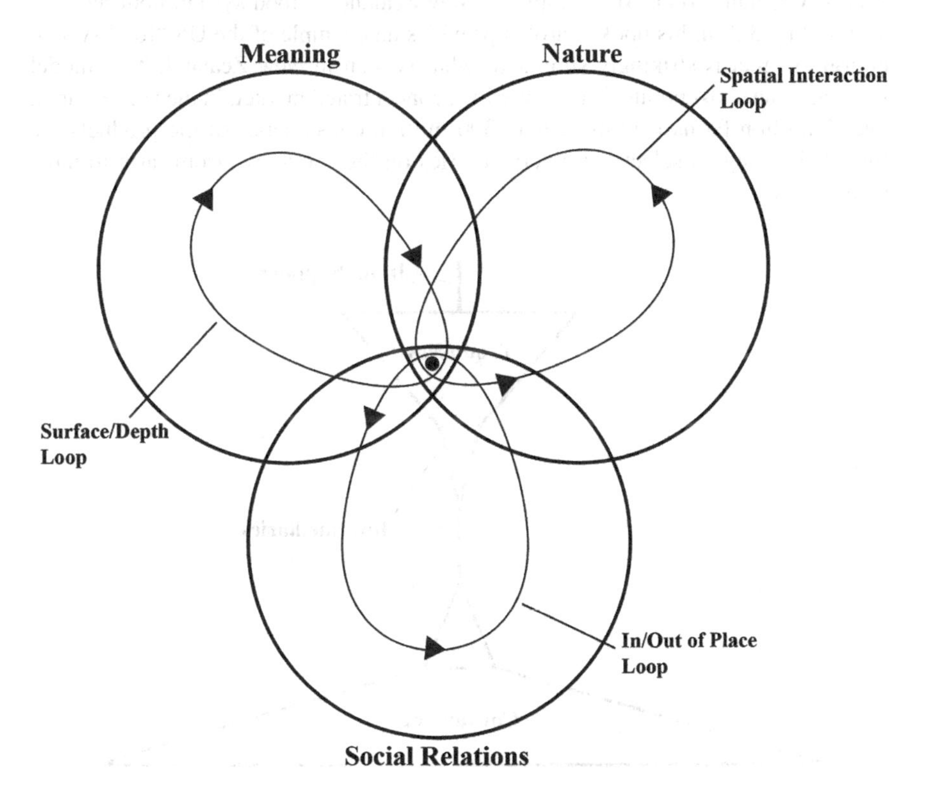

Fig. 3.2 Sack's conceptual diagram of "place as a force" (adapted from Sack 1997)

China. The GM advertisement shows a couple at a table in the middle of a field dominated by wild grass-cover. Their GMC Jimmy is seen in the background with mighty mountain peaks towering behind it. This allows the consumer to construct social and locational contexts. Owning a Jimmy makes the consumer an adventurer by allowing them to explore and understand the untouched wild landscape (Sack 1992). The Royal Doulton ad features a teacup and saucer set in a picturesque setting with a luxurious cabin and a tranquil lake. "One romantic setting deserves another" is written across the well-manicured grass. This ad suggests that owning fine china "empowers" the consumer to create a new setting—"one that differentiates [them] from other people" (Sack 1992).

3.3 Michael Carolan and Food System Concentration

The work of Carolan, a rural sociologist, can also be incorporated into a political–ecological lens with regard to yerba mate. In his 2012 book, Carolan introduces the "food system hourglass" to visualize market concentration in agricultural supply chains (Carolan 2012). An example of New Zealand's "food system hourglass" is seen in Fig. 3.3. In his book, Carolan provides an example of the US "food system hourglass" that is strikingly similar to what is seen in New Zealand. This model illustrates the chokepoints that occur with a concentrated market. In the US example, over 2 million farms feed more than 300 million consumers, but the products are funneled through a select few intermediaries on their journeys from farm to table (Carolan 2012).

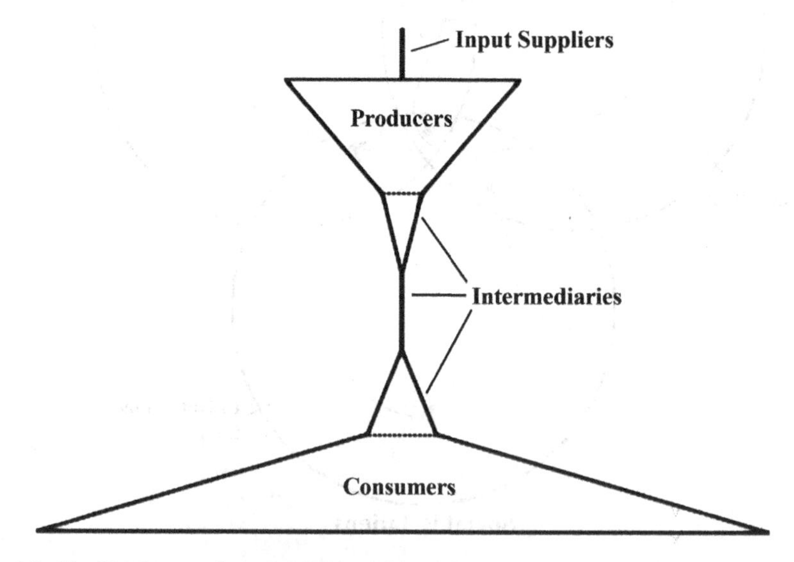

Fig. 3.3 The "food system hourglass" (adapted from Carolan 2012)

The existence of a strong buyer's market is difficult to avoid when sellers far outnumber buyers. Affording buyers increase market power as their numbers increasingly dwindle through corporate consolidation. Furthermore, the entire hourglass hangs by a thread because it represents the dozen or so dominant input suppliers, providing seeds, agrichemicals, and machinery (Carolan 2012). In this way, primary producers are squeezed from both ends, having very few choices for the purchase of and inputs and having relatively few choices of buyers for their outputs. As input suppliers raise prices or further consolidate, costs are forced onto primary producers who then sell to processors and manufacturers demanding as low a price as possible. Simply put, fewer large firms have much more negotiating power, resulting in a concentrated food system.

3.4 Putting it all Together: Political Ecological Lens Imbued by Sack, Cook, and Carolan

In the case of yerba mate, purely economic forces to keep costs down across the supply chain "take place," bringing these choices into play in the construction of place within the three realms, meaning, nature, and other sectors of social relations. The places generated by production are impacted by power across the supply chain, as illustrated with the ideas of Carolan.

Geographer Ian Cook's "Follow the Thing" methodology, as introduced in his 2003 article examining the papaya supply chain, serves as a particularly effective means of activating Sack's surface/depth loop, which again allows us to scratch at the surface and see the depth, or reality.

Cook's "Follow the Thing" methodology that has since become a dominant theme of his career. His approach employs the use of people who are "(un)knowingly connected to each other through the international trade in fresh papaya" to illustrate an "entangled range of economic, political, social, cultural, agricultural and other processes" (Cook et al. 2004). In this way, Cook engages with Harvey's "call for radical geographers to 'get behind the veil, the fetishism of the market,' to make powerful, important disturbing connections between Western consumers and distant strangers whose contributions to their lives were invisible, unnoticed, and largely unappreciated" (Cook et al. 2004). This work focuses mostly on consumption of a product from the Global South by residents of the same country, but it shows that the obscured nature of complex production chains can extend to products largely consumed in their country of origin.

Do the relationships and places constructed by consumption fulfill Sack's criteria of an intrinsically good place (increasing diversity and allowing people to see through to the real)? In our "thin," globalized world, too often the negative consequences of a consumer's economic choices (driven by consumption) are obscured. By combining ideas from across Sack's career and applying them to Cook's "Follow the Thing" approach, we can begin scratching away at the surface of the imagined places constructed through yerba mate advertising, and examine the true reality in hopes of building a more accurate understanding of distant place-making spurred by consumption, or in Harvey's words, to "get behind the veil" (Harvey 1990).

References

Blaikie PM (1985) The political economy of soil erosion in developing countries. Longman Development Studies. Longman, London, U.K.

Blaikie PM, Brookfield H (1987) Land Degradation and society. Methuen, London

Campbell DJ, Olson JM (1991) Framework for environment and development: The kite. In: CASID Occasional Papers. Michigan State University, East Lansing, MI, pp 1–30

Carolan M (2012) The sociology of food and agriculture. In: Hardwick T (ed) Earthscan food and agriculture. Taylor & Francis Group, London

Castree N, Kitchin R, Rogers A (2013) A dictionary of human geography. Oxford Paperback Reference, Oxford University Press, Oxford, U.K.

Cook IJ et al (2004) Follow the thing: papaya. Antipode 36(4):642–664

Gunn MC (1980) Cultural ecology: a brief overview. Neb Anthropol 5(149):19–27

Harvey D (1990) Between space and time: reflections on the geographical imagination. Ann Assoc Am Geogr 80(3):418–434

Perreault T, Watts MJ (2011) In: Hubbard P, Kitchin R (eds) Key thinkers on space and place. Sage, Los Angeles, CA

Robbins P (2012) Critical introductions to geography. In: Political ecology: a critical introduction, 2 edn. Wiley, Chichester, U.K.

Rosin CJ (2004) The political ecology of Mercosur/L: local knowledge and responses to a competitive market. University of Wisconsin-Madison

Sack R (1992) Place, modernity, and the consumer's world. The Johns Hopkins University Press, Baltimore.

Sack R (1997) Homo geographicus. Johns Hopkins University Press, Baltimore

Sack R (2001) The geographic problematic: empirical issues. Nor Geogr Tidsskr 55:107–116

Sack R (2003) A geographical guide to the real and the good. Routledge, London

Vayda AP (1983) Progressive contextualization: methods for research in human ecology. Hum Ecol 11(3):265–281

Vayda AP, Walters BB (1999) Against political ecology. Hum Ecol 27(1):167–179

Walker PA (2005) Political ecology: where is the ecology? Prog Hum Geogr 29(1):73–82

Zimmerer KS (2000) Rescaling irrigation in Latin America: the cultural images and political ecology of water resources. Cult Geograph 7(2):150–175

Zimmerer KS, Bassett TJ (eds) (2003) Political ecology: an integrative approach to geography and environment-development studies. Guilford Press, New York

Chapter 4
INYM, Prices, and the Argentine Yerba Mate Food System

Abstract This chapter details modern commercial yerba mate production in Argentina. By tracing yerba mate's supply chain, from seed to shopping cart, significant market distortions become apparent (particularly through consolidation and vertical integration of the larger firms). As the food system modernizes, small producers are squeezed on all ends—with decreasing fertility, increasing reliance on inputs, and a diminished share of the product's final purchase value.

Keywords Modernization · Food systems · Agricultural intensification · Market concentration

4.1 Current Socioeconomic Contexts: Argentina and Misiones

Misiones lags behind much of Argentina socioeconomically, making the province among one of the country's poorest. According to a 2016 report, 37.7% of Misiones' population lives in poverty, the highest of all Argentine provinces (IADER 2016). This figure places Misiones' poverty rate more than three times higher than what is seen in the city of Buenos Aires (IADER 2016). Due to currency devaluations since the report was issued, these figures are likely even higher now as wages lag behind consumer prices. Official statistics from INDEC suggest a recent decrease in poverty in Argentina, but poverty figures from INDEC are suspect due to manipulation for political reasons. In fact, from 2013 until 2016 the government (then led by populist President, and current Vice President, Cristina Fernández de Kirchner) stopped publishing these statistics after claiming a nationwide poverty rate of just 5%, lower than that of Germany (Squires 2018).

A more recent report published by UCA (Catholic University of Argentina) shows poverty among urban Argentines rising to 33.6%, an increase from just over 28% a year before (UCA 2018).

Considering that over 90% of the Argentine population is classified as urban, UCA's figures are especially striking (World Bank 2018).

At the time this research was conducted, 1 USD was equal to 40 Argentine Pesos (ARS) (Bloomberg 2019). One year ago, this exchange was 1 USD to 19.5 ARS,

A. S. Dohrenwend, *Green Gold*, SpringerBriefs in Latin American Studies,
https://doi.org/10.1007/978-3-030-82011-4_4

representing an over 100% drop in value over one year's time (Bloomberg 2019). In February of 2015, before the election of President Mauricio Macri, the exchange rate stood at roughly 1 USD to 8.5 ARS (Bloomberg 2019).

In 2018, consumer prices in the country rose 47.6%, the fifth highest increase globally and the second highest in Latin America, behind only Venezuela's more than 1,000,000% inflation rate (Otaolo and Garrison 2019). A number of concurrent factors explain Argentina's current economic problems. The major causes, however, relate to the national government's monetary policy, as well as the failure of much of 2018's soybean crop—an important source of hard currency due to its export value (Cohen 2018).

With respect to median monthly salary in the private formal sector, Misiones also ranks poorly among the other provinces and the city of Buenos Aires, ranking 23rd out of 24. Using 2012 exchange rates, Misiones's value of 4,342 ARS was roughly equal to 950 USD per month (IADER 2012). In terms of other socioeconomic indicators, Misiones consistently ranks near the bottom of Argentine provinces. The Human Development Index (HDI), published by the United Nations Development Programme, is a composite index out of 1.0 that takes into account life expectancy, education, and per capita income indicators. The province's HDI is 0.829, ranked 20th out of 24 (including the city of Buenos Aires) (UNDP 2018). The countrywide HDI is 0.848, while the city of Buenos Aires scores a 0.885 (UNDP 2018).

As the HDI does not account for inequality, a new metric called the Inequality-adjusted Human Development Index (IHDI) was introduced in 2010 (UNDP 2018). Most countries perform significantly lower in this index, as opposed to the more traditional HDI. Argentina's countrywide IHDI score is 0.707 (tied with Iran at 47th place out of the 151 country's with published calculations) (UNDP 2018).While this index is not available at the provincial scale, Misiones would likely place significantly lower.

4.2 The Argentine Yerba Mate Food System

Utilizing Carolan's model of visualization of concentrated systems provides a useful framework for examining Argentina's yerba mate food system. An adaptation of his model for this context is shown in Fig. 4.1. The CR4 model (four-firm concentration ratio) is a method to determine a level of concentration within a system. Utilizing the CR4 model to illustrate market concentration within a food system helps illustrate chokepoints along the supply chain. This supplements a political–ecological approach because the market concentration in the industrial sector allows for that sector to exert much stronger pressure on those placed earlier on the supply chain. The CR4 examines the combined market share of the four largest actors within a market and determines levels of concentration. Values of 20–39% indicate a concentrated market and values of 40–59% indicate a highly concentrated market (Carolan 2012).

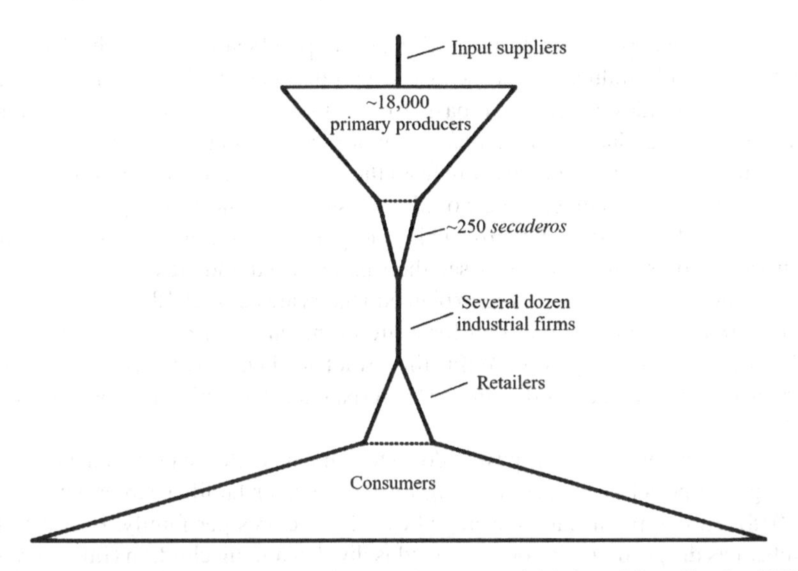

Fig. 4.1 The domestic yerba mate hourglass (not drawn to scale) (adapted from Carolan 2012)

As stated previously, the five largest firms in the yerba mate processing industry represent over 50% of the consumer market (Boerr 2018; Interviewee 3, 2018). The CR4 value calculated for 2017 utilizing INYM data is 48.7%, clearly surpassing the threshold for a highly concentrated market (Boerr 2018). According to these data, just one company alone rules 19.5% of the total consumer market in Argentina (Boerr 2018). *Las Marias,* by itself, is just below (by 0.5%) the consumer market share threshold to denote a concentrated market. The value for Las Marias was 21.1% for the previous year, surpassing the 20% threshold that year by itself (Boerr 2018).

The following figure depicts market concentration in Argentina's yerba mate sector, which illustrates how the phenomenon is similar to other commodities across the globe. With roughly 18,000 producers—of which the vast majority are small producers who maintain several hectares or less in production—there are around 250 driers to sell to.

The other complication lies in the increasing levels of vertical integration in the Argentine yerba mate industry. A significant amount of yerba mate producers grow on contract for one of the dominant companies involved in the supply chain (Interviewees 3, 4, & 5, 2018). This means that these companies, to varying degrees, are able to vertically integrate. These contracts can lead to further issues. Contracts are based on the value of the Argentine peso at the time they are signed, however, with today's volatile value of the peso, when the contract is paid out to the primary producer, sometimes months later, the value of the peso has often plummeted further (Interviewees 3 & 4, 2018). This can make the situation for small producers even more precarious.

Furthermore, over the past 30 years, concentration in land ownership has grown for companies (Interviewee 3, 2018). Land prices are generally very low in Misiones. A hectare sells for approximately 1,500 USD. Push–pull factors explain the decreasing number of small producers over the last several decades. *A Theory of Migration* by Lee (1966) provides a theoretical basis for push–pull factors. A *push* factor forces residents out of a place, while a *pull* factor attracts to a place. In the case of yerba mate, push factors push people away from the rural *yerbales*, while pull factors pull them towards urban centers like Posadas. Push factors lie in the poor conditions and quality of life for *yerbateros*, while the pull factors center around the allure of urban, modern life. "The sons see their parents, old and sick. They don't want that livelihood," one interviewee explained (Interviewee 4, 2018). The children of many *yerbateros* (producers) long for a life in the city, *"Ellos quieren WiFi, ellos quieren bailar"*—"They want WiFi, they want to dance" (Interviewee 3, 2018). Rural depopulation in areas dominated by yerba mate cultivation is shown in Gortari (2007).

Furthermore, when land is passed down to heirs, it is often subdivided, potentially making it impossible to make a living for the younger families (Interviewees 3 & 4, 2018). At one point, farms averaged over 20 hectares per family. However, this number has dropped precipitously as land is divided among children (Interviewee 7, 2018). People are not necessarily "forced" off of their land, but it is a "choice" that is influenced by the previously mentioned factors, among others. As producers vacate their land, "big companies can just buy; they have plenty of money" (Interviewee 4, 2018). These companies are not just limited to players in the yerba mate industry. The forestry industry in the region has grown immensely in recent years (Interviewee 4, 2018). Some farmers who remain on their land often change their livelihood in favor of other crops, such as tobacco and pine (Interviewees 3 & 4, 2018).

4.3 Pricing Regime

The most common complaint from producers is the low price paid for *hoja verde*. Figure 4.2 shows prices set by INYM's Board of Directors for yerba mate from 2002 (when INYM was founded) up to the current year. Prices for *hoja verde* and *hoja canchada* are set by the board biannually, in March and September. March prices take effect each April and conclude each October when they are replaced by the new negotiated price. Prices in ARS were compiled from INYM publications as well as from Roberto Montechiesi, a local agricultural engineer who is very influential in the yerba mate supply chain in Misiones. From there, the prices per kilogram in pesos for each category were converted to their equivalent in dollars during the year they were established. The two prices for *hoja verde* and *hoja seca* each year were averaged. The monthly ARS to USD exchange rates were also averaged for each year in order to obtain a single exchange value for each year. The average values for each stage of the yerba mate supply chain were multiplied by the corresponding average exchange rates yielding an average price per kilogram in US dollars for each year. In light of the change in mass caused by the drying process, both categories were examined by the value of their dry weight, meaning that to calculate the amount of money

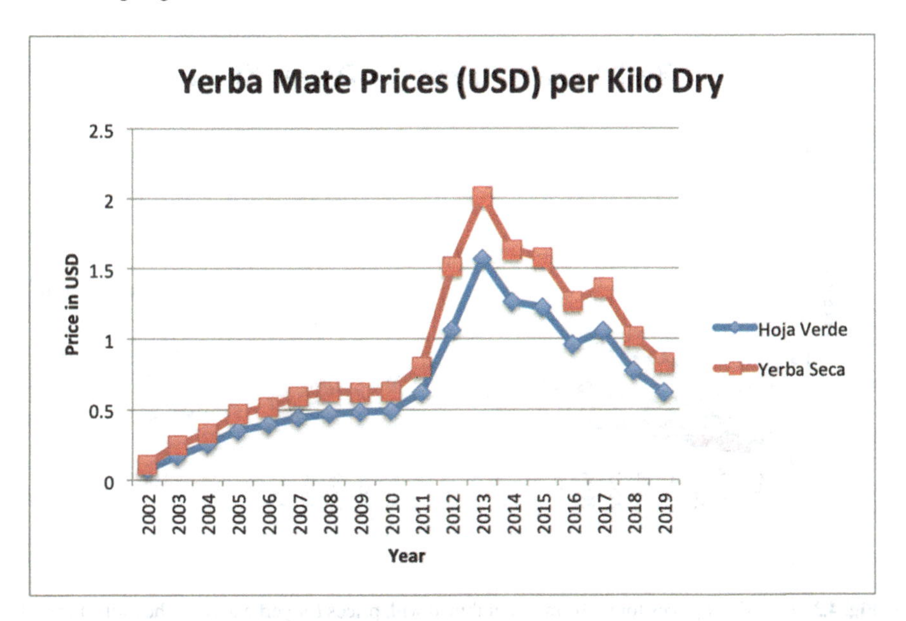

Fig. 4.2 INYM-set prices for yerba mate (adapted from Montechiesi 2016)

generated for a primary producer per kilogram of yerba mate, the per kilogram rate for *hoja verde* was multiplied by 2.857 (the average amount of kilograms of *hoja verde* required to yield one kilogram of *hoja canchada*).

The graph shows steady growth in price from 2002 (the year Argentina floated the peso) until about 2008, followed by several years of stagnation. Starting with the prices set for 2011, prices for both categories increased rapidly until 2013, when the *hoja verde* required to yield a kilogram of dried yerba mate was set at 1.57 USD. At this time, the per kilogram price paid to primary producers for *hoja verde* was 0.55 USD. The dry price per kilogram was 2.01 USD, meaning 0.44 USD of value was added by the drying process at the *secadero* as compared with the value of *hoja verde* required. This value remains with the drying firm, but about 0.10–0.15 USD is profit that remains after expenses such as fuel, labor, and infrastructure are accounted for (Montechiesi 2016) (Fig. 4.3).

The second graph expands on the first by adding in estimated prices of yerba mate once it leaves the mills and at retail establishments. These data are available from 2002 until 2015 from Montechiesi (2016). Though values were not available after 2015, these values were simulated by maintaining the 2015 differences in price between them and *hoja canchada*. When adding these published values, in addition to the projected values, it is clear that even with the downturn in prices since 2013, both the industrial sector and the retail sector have significantly increased the difference between what they sell for and what is paid to primary producers and *secaderos*. Though prices in USD have dropped as inflation rises, it appears that these sectors are more insulated from the crisis than other segments of the supply chain.

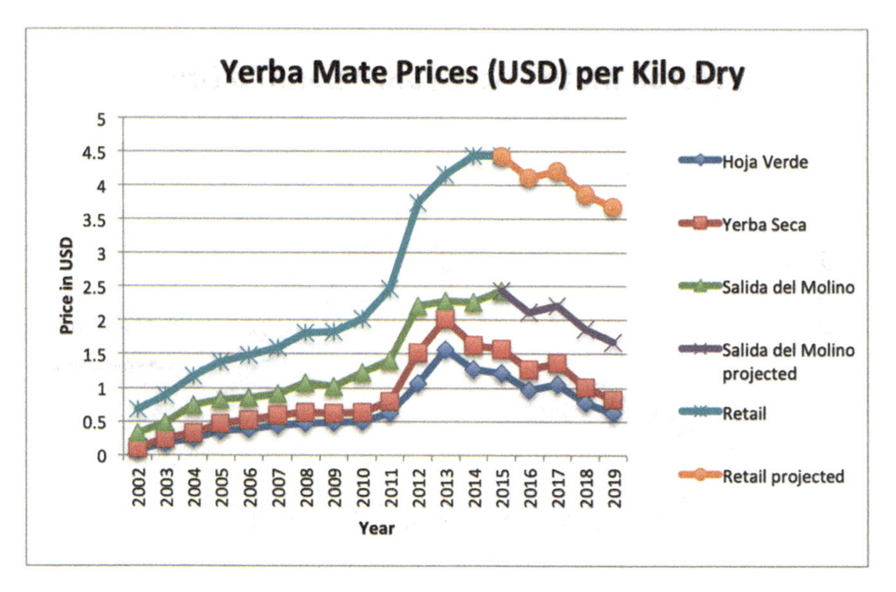

Fig. 4.3 INYM-set prices for yerba mate combined with prices for yerba leaving the mill and retail

A typical yield per hectare for a small producer averages around 4,000 kg per hectare, translates to a total revenue of 33,600 ARS at the current 8.4 ARS/kilogram price set by INYM (Interviewee 3 2018). At the current exchange rate, this equates to 840 USD. A typical smallholding is between 5 and 10 ha, meaning that in revenue, a smallholder might be expected to generate 4200–8400 USD per harvest over their entire holding. Though it is possible to yield as much as 12,000 kg of *hoja verde* per hectare with each harvest, this is out of reach for most for a few reasons. First, many of the *yerbales* are very old and feature quite low fertility after their many years in production. Education programs and extension projects are limited in the area, hampering training opportunities for small producers. Additionally, as small producers seek to harvest as much foliage as possible, overharvesting occurs frequently (Interviewee 7 2018). Overharvesting interferes with the tree's photosynthesis, limiting its ability to regenerate sufficiently for future harvests (Interviewee 7 2018). While yielding more money for the producer at first, overharvesting limits yields and revenue into the future for farmers among the most vulnerable in the yerba mate supply chain.

References

Bloomberg LP (2019) Usd-Ars X-Rate. https://www.bloomberg.com/quote/USDARS:CUR
Boerr M (2018) La Pelea Por El Mercado De La Yerba Mate: Las Marías Lidera, Pero Liebig No Para De Crecer Y Rosamonte Relegó a Molinos. Economis

Carolan M (2012) The sociology of food and agriculture. In: Hardwick T (ed) Earthscan food and agriculture. Taylor & Francis Group, London.
Cohen L (2018) Argentina's economic crisis explained in five charts. Reuters
Gortari J (ed) (2007) De La Tierra Sin Mal Al Tractorazo. Editorial Universitaria de la Universidad Nacional de Misiones, Posadas
IADER (2012) Santa Cruz, Chubut, T. del Fuego y Neuquén: pagan los sueldos más altos. Instituto Argentino para el Desarrollo de las Economías Regionales, Buenos Aires
IADER (2016) Vuelve a Alarmar La Pobreza: En Misiones, Chaco Y Formosa Alcanza Casi El 40%
Interviewee 3 (*Secadero*) (2018) Interview by A.S. Dohrenwend. Semi-structured interview. Posadas, July 2018
Interviewee 4 (Family of organic producers) (2018) Interview by A.S. Dohrenwend. Semi-structured interview. Oberá, July 2018
Interviewee 5 (Executive Director of large industrial firm) (2018) Interview by A.S. Dohrenwend. Semistructured interview. Posadas, July 2018
Interviewee 7 (Agricultural engineer) (2018) Interview by A.S. Dohrenwend. Semi-structured interview. Posadas, July 2018
Lee ES (1966) A theory of migration. Demography 3(1):47–57
Montechiesi R (2016) Yerba Mate De Ayer, De Hoy Y De Siempre. Posadas
Otaola J, Garrison C (2019) Argentine annual inflation hit 27-year high in 2018. Reuters
Squires S (2018) Argentina president says poverty to rise as economy slumps. Reuters
UCA (2018) Estancamiento Estructural, Pobrezas Crónicas, Exclusiones Económicas Y Desigualdades Sociales En La Argentina Urbana (2010–2018). Buenos Aires: El Observatorio de la Deuda Social Argentina, 2018
UNDP (2018) Inequality-adjusted human development index. United Nations, New York
World Bank (2018) Urban population (% of Total). World Bank, Washington DC

Chapter 5
Socio-Environmental Consequences of Low Margins in the Argentine Yerba Mate Food System

Abstract This chapter identifies and discusses the key social and environmental consequences of the modernizing food system outlined in Chap. 4. Based on a series of field interviews by the author (along with evidence from the literature), we see through yerba mate's careful branding to reveal the true contexts of the vast majority of production. Key issues explored include extreme poverty, child labor, occupational hazards, pollution, and soil erosion.

Keywords Yerba mate · Land concentration · Labor · Environmental degradation · Soil erosion · Agricultural intensification · Health

5.1 Land Concentration and Tenure

A significant issue in Misiones' agricultural landscape is the concentration of land in relatively few hands, which are often foreign entities. Neoliberal policy has aimed to attract foreign investment in the country by easing foreign land ownership regulations, among other strategies. The acquisition of nearly 1 million hectares of land in Patagonia by the Italian fashion brand Benetton illustrates these attempts. Benetton is among the largest landowners in all of Argentina, acquiring almost 1 million hectares in 1991 (Trouillet 2017).

This land was gifted to the British in 1889 after they helped finance the *Conquista del Desierto* several years earlier, and its ownership is heavily disputed by Indigenous activists in the region (Trouillet 2017). The military campaign virtually annihilated Argentina's Indigenous populations in Patagonia. In 1991, Benetton purchased the British firm with title to the land and became the new owner, which it now uses as an area to produce wool for their products (Trouillet 2017). Other major examples of controversial foreign land ownership in Argentina are seen in British ownership in Bariloche and Chilean ownership in Misiones's forestry industry (Interviewee 4 2018; Long 2009).

In 1881, the majority of Misiones province (2 million hectares) was sold to 29 investors, with one receiving a tract of 600,000 ha (Sawers 2000). The average tract size was 70,000 ha (Sawers 2000). Government programs in the early twentieth

A. S. Dohrenwend, *Green Gold*, SpringerBriefs in Latin American Studies,
https://doi.org/10.1007/978-3-030-82011-4_5

century alleviated the high concentration of land ownership by encouraging rural settlement and smallholder mate production (Sawers 2000).

More recent historical trends in ownership are characterized by the rise of large agribusiness holdings. According to the most recent agricultural census (2008), 30% of Misiones's agricultural land is held by corporations (INDEC 2008). Further analysis of INDEC data shows that each of the 132 corporate holdings is an average of 4,088 ha (INDEC 2008). When looking at the individual department level, inequalities become evident.

In the Iguazú department, a particularly sensitive area containing Iguazú Falls (one of the seven natural wonders of the world), there were 286,833 ha of agricultural land (INDEC 2008). Of these 262,865 (91.6%) were held by corporations and 266,502 (92.9%) were in holdings greater than 1000 ha (INDEC 2008). At the same time, the Guaraní department's 165,509 ha of agricultural land was only 8.6% corporate and 15.9% locked up in holdings of greater than 1,000 ha (INDEC 2008). The underlying reasons for this geographic disparity deserve further future attention.

Concerns surrounding the rise of corporate control do not necessarily suggest that smallholders were good stewards of the environment. As Sawers points out, since many smallholders do not have a clear title to their land, little incentive exists to provide adequate caretaking and to preserve fertility (2000). Though fertility loss can be remedied by fertilizer application, this would incur additional costs to smallholders. Smallholder cultivation is largely characterized by slash and burn practices that transform dense forests into barren land in as few as two years (Sawers 2000). This suggests that corporate land control is more so an issue of resource concentration specifically than of environmental degradation generally.

The vast majority of producers own the land they cultivate; however, many do not have a clear title to the land they own (Interviewee 3, 2018). Land typically sells for about 1,500 USD per hectare in the area (Interviewees 3 & 4, 2018). Upon purchase, many landowners obtain a *boleto de compraventa* (a bill of sale) or *permiso de ocupacion de las tierras* (an occupation permit) (Interviewee 3, 2018). The current structure of yerba mate production developed in 1926 when Argentine President Marcelo T. de Alvear's government offered provisional titles to farmers who planted between 25 and 50% of an area of land with yerba mate within two years of delivery (Rau 2009). This resulted in average increases in yerba mate cultivated area of about 5,000 ha per year over the following decade (Rau 2009).

5.2 The Plight of the Tareferos

Like most farmers, the *yerbateros* operate at incredibly challenging profit margins. Farmers must continuously seek methods to reduce costs in order to continue operating. These strategies to adapt to the ever-present pressure imposed by low margins can lead to serious social and environmental consequences. Low margins directly contribute to inadequate resources for hiring staff for field maintenance and harvest. They also diminish farmers' ability to purchase, maintain, and fuel field equipment like tractors. Routine maintenance includes, most significantly, "cleaning the field" (removal of weeds). It takes roughly 3–4 days to "clean" one hectare of a *yerbal* (yerba mate field) (Interviewee 3, 2018). For yerba mate, in particular, many producers use Roundup or other glyphosate-based herbicides in order to avoid the costs of cleaning the fields manually (Interviewees 3, 4, 6, & 7, 2018).

The primary weed species in the fields include both grasses and bushes. Grass species include *Poa pratensis* (bluegrass) and *Cynodon dactylon* (Bermuda grass); species of bushes include *Senna obtusifolia* (sicklepod), *Manihot esculenta* (manioc), and *Solanum granuloso-leprosum*, (*fumo bravo*, a species endemic to the area) (Interviewee 3, 2018). At around 20 USD per day, this represents 60–80 USD per hectare (Interviewee 4, 2018). With revenue of roughly 840 USD per hectare, this represents 7–10% of the total each time the field is cleaned. In addition to labor costs, producers also must pay for equipment and fuel, making the process of cleaning the fields even more expensive.

The harvest also represents a potentially significant expense for small producers. Often undertaken by low-paid, contract-less *tareferos* (those that harvest yerba mate), the job is backbreaking. Armed with hand clippers, *tareferos* begin work between 4 and 5am and continue until after dark (Pérez and Brito 2019). When fully loaded, the bags are extremely heavy and must be carried by hand if equipment is not available (Interviewee 3, 2018).

Wages are paid according to the amount of plant material harvested. The strongest *tareferos* are those able to collect and carry the most plant material (around 500 kg in one day), can expect typical wages between 150–200 USD per month (Pérez and Brito 2019). This is far below the province's median monthly salary. Many of these *tareferos*, who number in the thousands, work informally—migrating from between informal outdoor settlements under Misiones's dense tree cover (Ocampo and Kordi 2016; Pérez and Brito 2019). Photographs of a typical *tarefero* living situation can be seen in Pérez and Brito (2019).

5.3 Child Labor

In order to reduce costs, low margins often leave smallholders and their families responsible for these arduous tasks, unless they are willing to rely on informal labor among migrant families that greatly increase their costs of production. This has led to reliance on child labor in the industry, so much so that yerba mate is recognized on the United States Bureau of International Labor Affair's 2016 list of Goods Produced by Child Labor or Forced Labor (BILA 2016). A similar 2017 classification by the U.S. reported that children as young as five years old work in fields to help their parents harvest yerba mate– often by carrying heavy loads (BILA 2017). Child labor's pervasiveness in the industry extends to both the families of primary producers, as well as the families of migrant *tareferos*. Though large companies such as Las Marias claim to not use child labor on the land they cultivate themselves, many of their contracted suppliers do (Interviewees 1, 3, & 5, 2018).

Estimates show that roughly 16% of children in Misiones have never attended school, and instead mostly work in the agricultural sector (Vera 2017). Though it is possible for a smallholder to eke out an adequate living for their family, this requires effective management, which is often lacking, and significant effort by the entire family, which makes it difficult for children to receive an education (Interviewees 1 & 3, 2018).

In the past, child labor in the yerba mate harvest was largely unknown to most Argentines in distant Buenos Aires, Córdoba, Rosario, and Mendoza. A 2015 accident, however, spurred significant efforts to raise awareness of the dangers of child labor among activist groups. A truck carrying 17 child laborers who were destined for yerba mate fields overturned—killing Fernando, 13, Lucas, 14, and Edgard, 17—as well as Fernando's father who used his body as a human shield in an attempt to protect his son (Kordi 2014).

At the center of the campaign to end of child labor in yerba mate is human rights and literacy activist Patricia Ocampo, founder of *Un Sueño para Misiones* (a non governmental organization based in Posadas, the provincial capital) (pictured with the author in Fig. 5.1). Ocampo described the campaign in a 2017 interview with *La Nacion* as *"Es una campaña que nace del dolor y la muerte, ya que surge luego de que tres niños murieran al desbarrancase un camión en el que viajaban rumbo a un yerbal"*—"It is a campaign born from pain and death, since it arises after three died when the truck they were traveling to crashed on their way to a yerba mate plantation"

Fig. 5.1 Photograph of Patricia Ocampo with author, Posadas, July 2018 (photograph by author)

Fig. 5.2 "Me Gusta el Mate Sin Trabajo Infantil" Logo (Kordi 2014)

(Ayuso 2017). Ocampo later said, *"…es una situación que aún es desconocida por muchos Argentinos, por lo cual resulta fundamental seguir concientizando"*—"It is a situation that is unknown to many Argentines, so raising awareness is essential" (Ayuso 2017).

The NGO coined the phrase *"Me Gusta el Mate Sin Trabajo Infantil"*, Spanish for "I Like Mate Without Child Labor" (this is seen in Fig. 5.2). In 2015, the NGO, in cooperation with Posibl. Media (an Argentine media company), released a documentary film bearing the NGO's slogan as its title (Ocampo and Kordi 2016). The film, available for free on YouTube, documents the struggles of farmers and their families and was screened at the Cannes Film Festival in 2017 (Vera 2017). In addition to the documentary film, the NGO maintains a change.org petition that has received over 85,000 signatures as of 2019 and has been featured in media outlets like the BBC and CNN, which places its reach at an estimated 60 million people worldwide (Kordi 2014; Vera 2017).

The central goals of *Un Sueño para Misiones* have been to raise awareness and promote policy at the national and provincial levels by advocating for a bill that would create a certification system to help discourage the use of child labor in yerba mate cultivation and harvest. Though voluntary, companies that bear the certification on their packaging would be able to charge a higher price for the final product (Cámara de Diputados de la Nación Argentina 2017). The extra money would be allocated to an increase in wages for harvesters, with the understanding that without wage increases, the elimination of child labor would decrease family earnings (Cámara de Diputados de la Nación Argentina 2017). Without an increase in wages for harvesters, the bill would hurt the very children it is meant to help.

Ocampo's statements illustrate what Sack describes as the activation of the surface/depth loop, asdiscussed earlier. By raising awareness, she is scratching away at the surface of a landscape to reveal the consequences of consumption that are often obscured in our thinned-out [diffuse?] global economy. In her own words, *"Hoy no existe la trazabilidad – los procedimientos que permiten seguir el proceso de un producto en cada una de sus etapas"*, "Today, there is no traceability, the procedures that allow us to follow the process of a product in each one of its stages" (Ayuso 2017).

5.4 Health-Related Hazards Associated with Chemical Inputs

The use of chemical inputs in the fields, most commonly herbicides, is another example of a method to reduce costs. Glyphosate, the active ingredient in Roundup, developed by Monsanto (defunct since Bayer's 2018 acquisition), is the most commonly used herbicide in the world (Perry et al. 2019). Its use in Argentina is no exception. Between the years 1994 and 2010, the area of cultivated land grew by 45% and the amount of herbicide used increased from 19,376 metric tons of active ingredients to 227,185 metric tons, an increase of over 1000% (Livingstone 2016). More recent data show that Argentina now uses 240,000 metric tons of glyphosate (Avila-Vazquez et al. 2018). For yerba mate specifically, the majority of producers utilize Roundup or other glyphosate-based herbicides in order to avoid the costs of cleaning the fields manually as previously discussed (Interviewees 3, 4, 6, & 7, 2018). The ubiquity of glyphosate-based herbicides manifests itself in the contamination of the province's rivers—major sources of drinking water (Avigliano and Schenone 2015).

Glyphosate, a non selective, broad-spectrum herbicide absorbed mostly through foliage, is used to control unwanted broadleaf plants and grasses that compete with desired crop species (Henderson et al. 2010). The chemical interferes with the shikimate pathway, which is used by plants to biosynthesize critical amino acids (Henderson et al. 2010). In Argentina, glyphosate sells for around 5 USD per liter (not inclusive of government-imposed taxes) and, according to the manufacturer's guidelines, the standard rate of usage is 5 L per hectare (MERCOSUR 2017; Monsanto, n.d.).

In 2015, the World Health Organization's International Agency for Research on Cancer determined that glyphosate, along with some other agrochemicals, was "probably carcinogenic to humans" (IARC 2015). The organization acknowledged the evidence of carcinogenicity in humans for non-Hodgkin lymphoma shown in studies of exposures in agricultural settings from the United States, Canada, and Sweden (IARC 2015). Furthermore, strong evidence exists for the carcinogenicity of glyphosate in laboratory animals and is shown to cause DNA and chromosomal damage in human cells, with one study providing evidence of "increases in blood markers of chromosomal damage after glyphosate formulations were sprayed nearby" (IARC 2015).

Beyond the impact of the active ingredient, increased attention is being given to “inert” compounds in the final formulations of glyphosate-based herbicides. A compound in an agrochemical that is defined as “inert” is simply one that does not kill pests or weeds, regardless of toxicity (Gammon 2009). One such inert chemical is POEA (polyethoxylated tallow amine), which was found to be 3450 times more toxic than glyphosate itself (Defarge et al. 2018). Additionally, heavy metals such as arsenic, chromium, cobalt, lead, and nickel were found in assessed agrochemicals (including 11 glyphosate-based formulas) (Defarge et al. 2018). These toxic compounds are known endocrine disruptors and are shown to have impacts at levels below what is required for cytotoxicity (Defarge et al. 2018).

Glyphosate-based herbicides and their suspected health effects have been recently put on trial in the United States. Of the more than 11,000 lawsuits filed related to Roundup, only two have gone to trial (both in California) (BBC 2019). The first was in August 2018, when a state court jury concluded that Roundup caused a plaintiff’s cancer (BBC 2019). The second trial, held in San Francisco in March 2019, ruled unanimously in the same fashion (BBC 2019). Bayer’s stock price immediately plunged by 12% after the ruling (BBC 2019).

In Argentina, significant attention has been drawn recently to the impacts of glyphosate-based herbicides, especially after the publicization of the work of Dr. Damián Verzeñassi and his medical students at the National University of Rosario. Though not published in a scientific article, their findings received extensive publicity from domestic media (Ortiz 2017). Verzeñassi and his team conducted an epidemiological study in rural Argentina where herbicides are heavily applied and showed similar community health profiles that were uncharacteristic for Argentina as a whole (Ortiz 2017). Among the 26 communities examined, 80% of the total 87,382 residents lived fewer than 1,000 m from fumigation fields (Ortiz 2017).

On the national scale, most deaths in Argentina are caused by cardiovascular problems, however, in the communities examined, a variety of cancers were responsible for most deaths (Ortiz 2017). Beyond that, high rates of endocrine disorders (like hypothyroidism) were exhibited across the communities (Ortiz 2017). Nationally, 217 cases of cancer per 100,000 residents were recorded in 2012 (Ortiz 2017). In the communities examined, this figure was 397.4 (48.7% higher) (Ortiz 2017). An agrochemical lobbyist deflected when confronted with Verzeñassi’s findings by a French journalist. However, when asked by the journalist if he would be willing to drink a glass of the herbicide, the lobbyist responded, “I’m not stupid” (Ortiz 2017).

Another recent study examined the relationships between glyphosate exposure and reproductive issues in Monte Maíz, a typical Argentine agricultural village. 650 metric tons of glyphosate are applied per year in the area and glyphosate was present in village soil and in grain dust in the area (Avila-Vazquez et al. 2018). Concerns from medical professionals in the area about increasing rates of miscarriage and congenital defects spurred demands for the study (Avila-Vazquez et al. 2018).

The researchers found that miscarriages occurred at triple the national rate (10% vs. 3%) and congenital abnormalities occurred at double the national rate (4.3% vs. 1.4%) (Avila-Vazquez et al. 2018). The authors note that, though they demonstrated a correlation between glyphosate exposure and reproductive issues, further work is needed to demonstrate causation, similar to their findings in a previous study that demonstrates a geographic link between glyphosate exposure and increased frequencies of cancers in Argentine agricultural settings (2–3 times higher than the national rate) (Avila-Vazquez et al. 2017).

Perhaps the most striking quote from Verzeñassi in El Diario's profile is the following: "*¿Cuánto crecimiento de PIB de un país justifica la leucemia de un niño? ¿Cuánto crecimiento justifica un niño nacido con malformación, el desarrollo de cáncer, de hipotiroidismo en una persona? ¿Cuánto cuesta nuestra salud? ¿Quién y cuándo decidió que la vida se puede medir en términos económicos?*"—"How much growth of a country's GDP justifies a child's leukemia? How much growth justifies a child born with a malformation, the development of cancer, of hypothyroidism in a person, how much does our health cost? And when was it decided that life can be measured in economic terms?" (2017).

5.5 Environmental Harm Associated with Chemical Inputs

In addition to the potential impacts on human health, herbicide use also exacerbates soil erosion. Much of the hilly land receives persistent rainfall, thus, soil erosion is a significant issue even without taking herbicide use into account. The reasons for increased rates of soil erosion are due to the inability to maintain a system of cover-cropping when producers rely on broad-spectrum herbicides. Since the applied herbicide would kill a cover-crop in place, significant use of cover-cropping has become uncommon in yerba mate cultivation (Interviewees 3 & 4, 2018). Cover-cropping has numerous benefits, including the limitation of soil erosion, increased biodiversity, and improvements to soil fertility (SARE 2007).

With respect to soil erosion, there are several reasons why cover-cropping is helpful. First, roots help hold the soil, providing a natural defense from water erosion (Curell 2015). Second, above-ground vegetation cover limits wind erosion through friction by slowing wind speed (Curell 2015). Third, the above-ground growth provides a canopy that shields the ground from rainfall splatter that can displace soil (Curell 2015). In yerba mate cultivation, research by Casas et al. (1983) suggests the loss of 25 metric tons of soil per hectare each year under monoculture operations. This means that on a plantation of 10 ha, 12,500 metric tons of soil would be lost over the course of 50 years of yerba mate monoculture production.

In addition to issues of soil erosion, the loss of fertility in the old *yerbales* deserves attention. The highly weathered soils of the tropical periphery are not fertile to begin with, and quickly lose what little organic matter they have (Piccolo et al. 2008). In order to demonstrate the degradation caused by yerba mate plantations, Piccolo et al. collected and analyzed soil from a yerba mate monoculture, a field crossplanted with *Pennisetum purpureum* (elephant grass), and from a typical natural forest in the area (Piccolo et al. 2008). The monoculture soil's C content was one-third that of the forest while its N content was one-half (Piccolo et al. 2008). While the use of elephant grass to limit soil degradation and erosion is promising, this practice is largely avoided by the industry (Piccolo et al. 2008).

Dealing with declining fertility is critical for improving the livelihoods of small producers. If a nitrogen-fixing cover crop is not possible, producers may turn to fertilizers. Though commercial fertilizer use is not as widespread as herbicide use, applying fertilizer to increase yields incurs another cost for small producers. Declining fertility in the *yerbales* reinforces rural poverty among small producers. Agroforestry techniques, while promising, have not been widely adopted by the industry (Ilany et al. 2010).

An organic movement does exist in the Argentine yerba mate industry, but it is not widespread due to several key challenges faced by organic producers. First, in order for a producer to certify their yerba mate as organic, inspectors from OIA in Buenos Aires (*Organización Internacional Agropecuaria*) must travel to the distant places of production, which contributes to the high costs of certification (Interviewee 4 2018). Second, due to the nature of organic production, organic yerba mate must either be processed in dedicated organic-only drying facilities or be processed at the start of the season, otherwise the organic yerba mate will be contaminated by drying facilities that have already processed non organic yerba mate that season (Interviewee 4 2018). Additionally, prices for different grades of yerba mate are not established by INYM, meaning that if organic yerba mate is sold into the industrial pipeline, higher prices to compensate for the potentially higher costs of production are not guaranteed (Interviewee 4 2018). This encourages organic producers to vertically integrate, meaning that the primary producer oversees the full production process for their yerba mate. This degree of vertical integration is often too organizationally daunting and financially out-of-reach for small producers.

References

Avigliano E, Schenone NF (2015) Human health risk assessment and environmental distribution of trace elements, glyphosate, fecal coliform and total coliform in Atlantic Rainforest Mountain Rivers (South America). Microchem J 122:149–158

Avila-Vazquez M, Maturano E, Etchegoyen A, Difilippo FS, MacLean B (2017) Association between cancer and environmental exposure to glyphosate. Int J Clin Med 8(2):73–85

Avila-Vazquez M, Difilippo FS, MacLean B, Maturano E, Etchegoyen A (2018) Environmental exposure to glyphosate and reproductive health impacts in agricultural population of Argentina. J Environ Prot 9(3):241–253

Ayuso M (2017) Una Campaña Busca Visibilizar La Situación De Extrema Pobreza Y Trabajo Infantil Que Se Vive En Torno a La Cosecha De La Yerba Mate. La Nacion

BBC (2019) Weedkiller glyphosate a 'Substantial' cancer factor. BBC News

BILA (2016) List of goods produced by child labor or forced labor. Untited States Department of Labor, Washington D.C.

BILA (2017) Child labor and forced labor report: Argentina. United States Department of Labor, Washington D.C.

Cámara de Diputados de la Nación Argentina (2017) Certificación De Productos Libres De Trabajo Infantil. 2795-D-2017

Casas R, Michelena R, LaCorte S (1983) Relevamiento De Propiedades Físicas Y Químicas De Suelos Sometidos a Distintos Usos En El Sur De Misiones Y Ne De Corrientes. 18 Buenos Aires: Instituto Nacional de Tecnología Agropecuaria
Cataltaneo MB (2018) Un Sueño para Misiones, Posadas
Curell C (2015) Controlling soil erosion with cover crops. In: MSU extention. Michigan State University, East Lansing
Defarge N, Spiroux de Vendômois J, Séralini GE (2018) Toxicity of formulants and heavy metals in glyphosate-based herbicides and other pesticides. Toxicol Rep 5:156–163
Gammon C (2009) Weed-whacking herbicide proves deadly to human cells. Sci Am
Henderson AM, Gervais JA, Luukinen B, Buhl K, Stone D (2010) Glyphosate general fact sheet.Oregon State University Extension Services, National Pesticide Information Center
IARC (2015) IARC monographs volume 122: evaluation of five organophosphate insecticides and herbicides. International Agency for Research on Cancer, Lyon
Ilany T, Ashton MS, Montagnini F, Martinez C (2010) Using agroforestry to improve soil fertility: effects of intercropping on ilex paraguariensis (yerba mate) plantations with Araucaria angustifolia. Agrofor Syst 80(3):399–409
INDEC (2008) Provincia De Misiones, Censo Nacional Agropecuario 2008. Instituto Nacional de Estadística y Censos, Buenos Aires
Interviewee 1 (Patricia Ocampo) (2018) Interview by A.S. Dohrenwend. Semi-structured interview. Posadas, July 2018
Interviewee 3 (*Secadero*) (2018) Interview by A.S. Dohrenwend. Semi-structured interview. Posadas, July 2018
Interviewee 4 (Family of organic producers) (2018) Interview by A.S. Dohrenwend. Semi structured interview. Oberá, July 2018
Interviewee 5 (Executive Director of large industrial firm) (2018) Interview by A.S. Dohrenwend. Semi-structured interview. Posadas, July 2018
Interviewee 6 (Large producer) (2018) Interview by A.S. Dohrenwend. Semi-structured interview. Posadas, July 2018
Interviewee 7 (Agricultural engineer) (2018) Interview by A.S. Dohrenwend. Semi-structured interview. Posadas, July 2018
Kordi J (2014) Eligimos Productor Libres De Trabajo Infantil.Change.org, PBC. https://www.change.org/p/dante-sica-monzoemilio-yerba-mate-sin-trabajoinfantil
Livingstone G (2016) The villagers who fear herbicides. BBC
Long G (2009) Chile's forestry industry. AmCham Chile, Santiago
MERCOSUR (2017) Glifosato Roundup Full Ii. MERCOSUR, https://www.mercosur.com/es/precio-de-glifosato.roundup.full.ii/
Monsanto (n.d) Sprayers and water volumes. Monsanto, https://www.monsantoag.co.uk/roundup/roundup-amenity/application-information/
Ocampo P, Kordi J (2016) Me Gusta El Mate Sin Trabajo Infantil. 30 minutes: Posibl. Media
Ortiz A (2017) Los Efectos Del Herbicida Glifosato En Argentina: "¿Cuánto Crecimiento Del Pib Justifica El Cáncer?" El Diario
Pérez A, Brito ML (2019) Tareferos: La Historia Detrás De La Yerba. Agencia ZUR
Perry ED, Hennessy DA, Moschini G (2019) Product concentration and usage: behavioral effects in the glyphosate market. J Econ Behav Organ 158:543–559
Piccolo GA, Andriulo AE, Mary B (2008) Changes in soil organic matter under different land management in Misiones Province (Argentina). Scientia Agricola 65(3):290–297
Rau V (2009) La Yerba Mate En Misiones (Argentina). Estructura Y Significados De Una Producción Localizada. Agroalimentaria 15(28):49–58
Sawers L (2000) Income distribution and environmental degradation in the argentine interior. Latin Am Res Rev 35(2):3–33
SARE (2007) Managing cover crops profitably. University of Maryland, College Park
Trouillet C (2017) En Patagonie, Les Indiens Relancent Leur Lutte Sans Fin Contre Benetton. Le Temps
Vera V (2017) El Documental Que Denuncia El Trabajo Infantil Detrás Del Mate Llegó a Cannes Y Emocionó Al Mundo. La Nacion

Chapter 6
Conclusions

Abstract This chapter outlines plans for fruitful further work, acknowledges study limitations, and provides concluding thoughts. Yerba mate's marketing in the Global North, far from its growing range in South America's Atlantic Rainforest, provides a constructed meaning to spur increased consumption. While true of yerba mate, similar examples are seen across the global commodity landscape. By utilizing a political-ecological lens like outlined in this text, researchers can better illustrate the often-distant and obscured relationships of global capitalist development.

Keywords Political ecology · Robert Sack · Yerba mate · Globalization

6.1 Future Work and Limitations

This work raises many unanswered questions, particularly during the fieldwork period during July 2018. Perhaps the most significant involves the growing trend of deforestation in Misiones. Previously, I expected yerba mate plantations to be rapidly encroaching on the region's remaining pristine rainforest. This was not the case, as confirmed through the interviews conducted with individuals across the production chain (Interviewees 1–7, 2018). The area dedicated to the cultivation of yerba mate has been in place for decades and has remained relatively stable for decades (Interviewees 2, 3, 4, 5, 6, and 7, 2018).

This does not mean deforestation did not previously occur in Misiones, however. The culprit appears to be the cultivation of pine and eucalyptus in the region for paper production, which is similar to what Chilean interests had done to their own temperate forests before turning to Argentina's Atlantic Rainforest. During drives and bus rides across the area, the landscape appeared to be dominated by vast tree plantations. Several interview participants cited the growing influence of multinational firms as a major and growing threat to the last significant expanse of South America's Atlantic Rainforest (Interviewees 3, 4, and 7, 2018). This issue is introduced in the discussion of land concentration in Misiones, despite the statistics not specifically relating to yerba mate cultivation.

The most significant limitation faced was a lack of cooperation from INYM after fieldwork was conducted. Several pieces of critical data were requested, which INYM

A. S. Dohrenwend, *Green Gold*, SpringerBriefs in Latin American Studies,
https://doi.org/10.1007/978-3-030-82011-4_6

declined to provide, citing producer privacy, and in spite of the fact that no personal information was requested and the stated goal was simply to examine general trends in land ownership and production statistics. Rather than just showing the realities of production as they are now, this data would have provided a much clearer quantitative picture of how the production chain and its power distributions have been changing over the last few decades.

As these activities do not occur in isolation from one another, fruitful future work could involve the examination of relationships between production and land-use choices among the area's rural population. If yerba mate does not hold financial promise, perhaps producers may turn to other cultivars, or even sell their land to a large firm whose activities may end up being even more socially and environmentally destructive.

In addition, the narrow scope of this work reflects the majority of yerba mate produced in Argentina, but not all of it. Though companies like Guayakí may have merit in some of their advertised claims in the Global North, this work does not seek to specifically examine them. Rather, the constructed meaning of yerba mate from this marketing was assessed in terms of the majority of Argentine yerba mate. Future work could involve a closer assessment of meanings of yerba mate and their assessment's application to products destined for the Global North. This work could examine production specifically for the export market and a discussion of both Brazilian and Paraguayan involvement would be incorporated.

6.2 Closing Thoughts

The global yerba mate industry is valued at 1.339 billion USD and is projected to grow to 1.65 billion USD by 2025 (Value Market Research 2019). As this growth occurs, several questions must be addressed. Is yerba mate a friend or foe of the environment in which it grows? What about those who grow it? How will this growth exacerbate current trends and how will this additional future valuation be distributed across yerba mate's production chain?

Before drawing too many conclusions, it is important to harken back to Sack's first sentence in *Homo Geographicus*: "We humans are geographical beings transforming the earth and making it into a home, and that transformed world affects who we are" (Sack 1997). Again, humans are, by nature, geographical actors. This means that the choices consumers make are in a constant and dynamic interplay with both social and the earth's biogeophysical processes and realities across multiple spatial scales, standing true with Zimmerer's definition of geographical political ecology (Zimmerer 2000).

Though the focus on the externalization of social and environmental costs is not novel and the phenomenon is not unique to yerba mate, this work serves as a case study through which several theoretical lenses can be combined to form a unique framework to better understand the consequences of consumption that often occur large distances away from the place of final consumption. The newly formulated lens outlined in this thesis can be applied to other consumer products in a similar fashion. Sack and Cook each provide powerful standalone frameworks, but the combination of the two allows for a more holistic examination of modern consumption and a quite-stinging assessment of the often far-flung places of production constructed as a result. This case study, along with numerous other political-ecological examinations of commodities, is consistent with Sack's geographic view of modernity.

Under a capitalist system that maximizes profit by externalizing cost, the case of yerba mate is consistent with Sack's theory. Hyperspecialized places of both production and consumption have become so thinned out, the only way to connect them is through marketing. Yerba mate is one of many "new things" under capitalism that is spreading to the Global North, following the lead of other plants, like quinoa and açai, with long indigenous histories in their native growing regions deep in South America. In fact, these indigenous histories help provide a basis for the marketed contexts in the Global North. The distant geographies of these products lend an exotic flair, however, there is nothing "new" about them other than their continued inclusion into the globalized capitalist market. This phenomenon spurs consumer interest in order to create demand through the creation of context. While these parts of yerba mate's story in South America are amplified by the globalized capitalist market, the consequences of consumption manifesting in the region are often quietly ignored.

Though Sack's work lends itself more obviously to the most globalized spatial commodity flows, it also fits well within the domestic yerba mate in Argentina. Ignoring the global aspect, for a moment, Patricia Ocampo's experiences in child labor awareness and activism are evidence that these processes can occur even without a globalized economy, solely in trade across Argentina, as illuminated from her statements discussed previously. The "thinned-out" places of urban landscapes like in Buenos Aires, Córdoba, Mendoza, and Rosario have little to no understanding of their consequences on rural environments and peoples in Misiones and Corrientes almost one thousand kilometers away.

Furthermore, Patricia Ocampo's work can be assessed by Sack's inextricably linked theoretical assertions. Despite having never read Sack, Ocampo's quotes sound as if they were plucked directly from the pages of a Spanish translation of *Homo Geographicus*. Under the prevailing circumstances, Ocampo's goal is to generate awareness among consumers, chiefly in Argentina, where the vast majority of Argentina yerba mate consumption is occurring. The recent proposals that have made it in front of Argentina's national legislature provide an example. By creating a certification system against child labor and affixing seals to in-compliance yerba mate packages, though voluntary, consumers will have more of a connection with places of production.

Turning to INYM, a complete discussion must acknowledge that the situation was even more precarious than before it was established. Extremely volatile pricing regimes before INYM's oversight contributed to an unstable boom-bust cycle in which farmers were at the mercy of the global market and those at the higher links of the production chain. Despite its imperfections, INYM's intended collaborative nature provides a seat at the table for small producers that might not have existed otherwise.

Under Sack's framework, consumers cannot make truly autonomous decisions without awareness, so a certification system could help address the hyperspecialized nature of both the places of production and consumption. If consumers purchase something that helps build more ethical places of production, they are also making a conscious decision against purchasing the alternatives, which do not help build more just places of production. Though the voluntary nature of the proposal may seem imperfect on its surface, the existence of a voluntary system could help amplify the differences between mates produced with and without child labor, creating an area for discussion and reflection among consumers that would also be consistent with Sack's theory.

The way in which yerba mate is defined by the global market does not reflect the realities of the vast majority of yerba mate production in Argentina. The nature of the production chain spurs vast social inequities that, as seen by the involvement of children, reinforce themselves over time. Furthermore, as costs are externalized on the environment, the ecological underpinnings of the industry begin to fray, which may threaten the ability of farmers to make an adequate living in the future.

Yerba mate's manifestation on the global stage checks numerous boxes for instrumental "goodness" under capitalism, as theorized by Sack. Yerba mate can be "bought low" and "sold high," generating profit along the supply chain. Yerba mate can also inspire a feeling of "togetherness" and benefit the health of its consumers. However, if profits, feelings of "togetherness," and health benefits are built on obscured social injustice and leukemia diagnoses in distant places, they are not consistent with Sack's theorized criteria of intrinsic morality.

Today's global commodity flows and marketing strategies can make these questions difficult for the average consumer to answer thoughtfully. Consumption sets in motion a vast network of social and environmental relationships that often span global contexts. Modernity comes at a cost, which is borne by the children, impoverished laborers, and the environment of Argentina's Atlantic Rainforest. By utilizing a political-ecological lens imbued by the work of Robert Sack, this work aims to "Follow the Thing" in order to address questions like these and add to the body of work on commodities and the consequences of consumption in our globalized economy.

References

Interviewee 1 (Patricia Ocampo) (2018) Interview by A.S. Dohrenwend. Semi-structured interview. Posadas, July 2018

Interviewee 2 (Marketing Manager) (2018) Interview by A.S. Dohrenwend. Semi-structured interview. Posadas, July 2018

Interviewee 3 (*Secadero*) (2018) Interview by A.S. Dohrenwend. Semi-structured interview. Posadas, July 2018

Interviewee 4 (Family of organic producers) (2018) Interview by A.S. Dohrenwend. Semi-structured interview. Oberá, July 2018

Interviewee 5 (Executive Director of large industrial firm) (2018) Interview by A.S. Dohrenwend. Semi-structured interview. Posadas, July 2018

Interviewee 6 (Large producer) (2018) Interview by A.S. Dohrenwend. Semi-structured interview. Posadas, July 2018

Interviewee 7 (Agricultural engineer) (2018) Interview by A.S. Dohrenwend. Semi-structured interview. Posadas, July 2018

Sack R (1997) Homo Geographicus. Johns Hopkins University Press, Baltimore

Value Market Research (2019) Global Yerba Mate Market Report by Form (Liquid Concentrate, Powder and Others), Application (Food and Beverage, Dietary Supplement, Cosmetic & Personal Care, Functional Food and Others), End-Use (Retail and Industrial) and by Regions-Industry Trends, Size, Share, Growth, Estimation and Forecast, 2018–2025. Value Market Research

Zimmerer KS (2000) Rescaling irrigation in Latin America: the cultural images and political ecology of water resources. Cultural Geograp 7(2):150–175

9783030820107